THE LAST PICTURES

THE LAST

PICTURES

TREVOR PAGLEN

CREATIVE TIME BOOKS   New York

UNIVERSITY OF CALIFORNIA PRESS   Berkeley Los Angeles London

University of California Press, one of the most distinguished university presses in the United States, enriches lives around the world by advancing scholarship in the humanities, social sciences, and natural sciences. Its activities are supported by the UC Press Foundation and by philanthropic contributions from individuals and institutions. For more information, visit www.ucpress.edu.

University of California Press
Berkeley and Los Angeles, California

University of California Press, Ltd.
London, England

Creative Time Books is the publishing arm of Creative Time, Inc., a public arts organization that has been commissioning adventurous public art in New York City and beyond since 1972.

Creative Time Books
59 East 4th Street, 6th floor
New York, NY 10003
www.creativetime.org

Library of Congress Cataloging-in-Publication Data
Paglen, Trevor.
   The last pictures / Trevor Paglen, Creative Time, New York.
      pages   cm
   Includes bibliographical references.
   ISBN 978-0-520-27500-3 (alk. paper)
   1. Paglen, Trevor—Themes, motives.   2. Interstellar communication.
I. Creative Time, Inc.   II. Title.
   N6537.P22A4   2012
   709.2—dc23   2012017150

Designer: Lia Tjandra
Compositor: BookMatters
Text: Leitura Sans
Display: Lato Light
Printer and binder: Thomson-Shore, Inc.

Manufactured in the United States of America

21   20   19   18   17   16   15   14   13   12
10   9   8   7   6   5   4   3   2   1

The paper used in this publication meets the minimum requirements of ANSI/NISO Z39.48-1992 (R 2002) (*Permanence of Paper*).

WITHDRAWN

CONTENTS

Left: Echostar XVI
Right: Artifact onboard EchoStar XVI

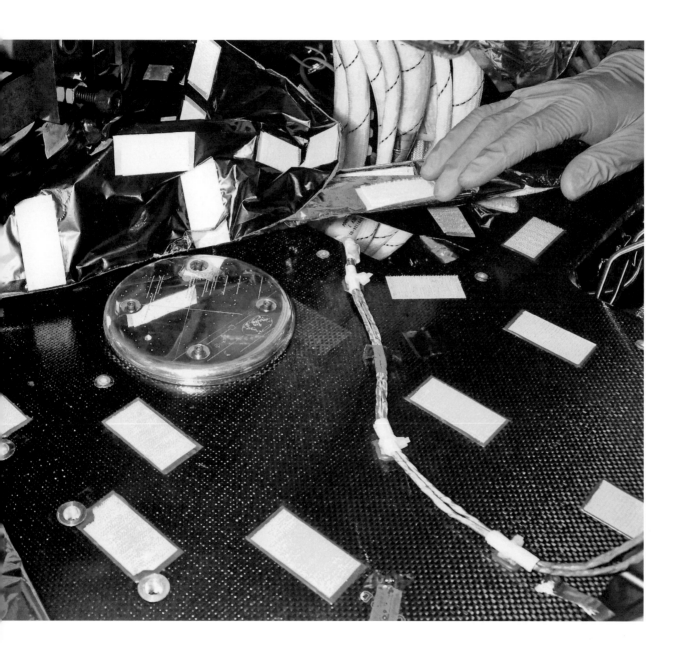

FOREWORD

When Trevor Paglen approached Creative Time with his dream of creating an artwork destined for the geosynchronous orbit, we jumped at the opportunity. For forty years, Creative Time has been presenting artists' dream projects in the public realm. For the past dozen years, our own dreams have included commissioning a major public art project in outer space. Like Captain James T. Kirk of the starship *Enterprise*—who famously declared, "Space: the final frontier"—we were seduced by Paglen's vision of attaching a time capsule to the side of a satellite heading to space. Creative Time is known for innovating artistic practice and adventuring into unknown territory. Certainly space represents what is perhaps the most unknown and mysterious place of all. But even more important than our excitement for space is the enduring purpose of this artwork in raising essential questions about the very colonialist desire that space has so often represented. After four years in development, Trevor Paglen has created *The Last Pictures*, a subtle and deeply thoughtful project to stir the collective imagination.

As much as this project was an opportunity for Creative Time to grow its programming beyond New York and cities around the world, it is an important opportunity for Trevor Paglen to push his artistic practice into new realms. Trevor is known for his stunning photographs that make visible the invisible. Using a lens for photographing distant stars and turning it horizontally across the land, he has revealed clandestine landscapes of the United States Defense Department. Turning that camera lens up toward the night sky, he has focused on the distant orbits of communications and spy satellites. Suddenly the earth and the sky above are filled not just with the mysteries and magic of nature, but also with the complex interventions of humankind. His inquiries continue anew with *The Last Pictures*, as Paglen inverts his formula by literally situating photographs he's already taken into new photos. Here the geography is that of the Clarke Belt, a geostationary orbit (GSO) teeming with thousands of spacecraft whose job it is to orbit the earth and broadcast television signals, route our telephone calls, and process credit card transactions. By dint of the profound stability of this orbit, these objects will outlive human civilization by billions of years—lasting longer than multicelled organisms have been on earth. It is a timescale difficult to imagine. If our oldest cave paintings are thirty-five thousand years old, what can we make of these series of representations circling the earth, eons from now, as an ancient representation of what was once our present?

The Last Pictures can be seen in the space tradition of fiction writer—and Trevor Paglen influence—Stanisław Lem. In the Russian tradition of cosmonaut literature, space is often a psychological arena that reveals more about society—its desires, frustrations, naïveté, and vanity—than about extraterrestrial life. Expanding on this lens, Paglen took up other creative confrontations with notions of deep time, from Carl Sagan and company's Golden Records attached to the unmanned space probes *Voyager 1* and *Voyager 2* in 1977, to an ongoing project to communicate the dangers of nuclear waste to denizens of the far future. To consider the possibility of communicating across space and time, Paglen set out to interview scientists, philosophers, anthropologists, archaeologists, mathematicians, and artists about the conflicts of modern society. He conducted a graduate seminar, with students researching images and debating the possibilities—and impossibilities—of communicating who and what we are today. Paglen participated in a residency at MIT's List Visual Arts Center and Visiting Artists program, where scientists worked with him on the production of these images in an archival format that would last millions of years. Energies were focused on solving just how we would actually launch the artwork into outer space. Calls and emails went out daily to satellite companies, launch brokers, scientists, and telecommunication companies. The task was not simple, but we were blessed to secure the partnership of EchoStar Corporation in November 2011.

Attached to the *EchoStar XVI* satellite is a golden disk approximately five inches in diameter containing a silicon wafer etched with one hundred images that form the core of this book and of the project itself. Paglen selected these images for their specificity, yet the intention of many of the images may not be self-evident. From an electron microscopic photograph of a Martian meteorite to the compelling *Family of Man* photograph of Yvonne Chevallier's 1952 trial, many of these images are opaque until their stories are told. And yet, collectively, *The Last Pictures* retain both a poetic and, dare we say, prophetic value. They are warnings. Their composition enacts a paradox worthy of civilization's epitaph, one that desperately communicates to the future the dangers of a highly capable and creative society unchecked.

With a project as ambitious as this, there are many people to thank. First and foremost, we want to thank our partners at EchoStar Corporation, Dish Network, The MIT List Visual Arts Center and Visiting Artists program, UC Press, and the Creative Time staff, volunteers, and researchers (see Acknowledgments). Most of all, it is an honor to have worked with Trevor Paglen. It is not often that one has the opportunity to send an artwork to space, and it is even more rare to meet someone with the will and courage to imagine and carry out a project as thoughtful and poetic as *The Last Pictures*.

Anne Pasternak
President and Artistic Director, Creative Time

Nato Thompson
Chief Curator, Creative Time

INTRODUCTION: GEOGRAPHIES OF TIME

In the fall of 2012, planet Earth will acquire a new moon. It won't be anything particularly remarkable. In fact, hardly anyone will notice. Over the last half century, since the former Soviet Union placed the first "artificial moon" into a low orbit on October 4, 1957, the world's nations and corporations have sent thousands of satellites spinning around the earth. Some of them monitor the weather; others take pictures or relay telephone calls. One of them, the International Space Station, even serves as temporary home for humans. In the early twenty-first century, there's nothing remarkable about adding new moons to earth's vicinity.

EchoStar XVI is a communications satellite. It doesn't do anything appreciably different from what *EchoStars I* through *XV* have done. Nonetheless, there's something unique about *EchoStar XVI*. Like its predecessors, *Echo-Star XVI* is designed to broadcast pictures on hundreds of television channels. But this satellite holds other pictures. Attached to its anti-earth deck, a gold-plated aluminum housing contains a small silicon disc designed to last for billions of years. This disc holds a collection of one hundred pictures.

EchoStar XVI and the image disc onboard have a profound and counterintuitive set of relationships to time. The satellite is an instrument of speed, transmitting hundreds of thousands of images per minute, all at the speed of light. Yet the future of *EchoStar XVI* lies in the deep time of the cosmos.

EchoStar XVI is a communication satellite, a technology of commerce, media, and global connectedness. It comes from a lineage of communication and transportation technologies that have, in a relatively short period of time, fundamentally altered our relationship to time and space. In the nineteenth century, with the advent of steam engines, trains, and telegraphs, the world became dramatically smaller. Railroads meant journeys that had once taken days or weeks were reduced to mere hours. Telegraph cables meant messages that were once hand-carried by men riding horses could now travel at the speed of light. Transportation and communication technologies integrated global commodities markets and led to the creation of international financial markets. Over the course of a single generation, time sped up so quickly and so dramatically that humans found it nearly impossible to keep up.

Artists and philosophers tried to make sense of a new world of speed, commerce, and distorted geographies. In 1844, J.M.W. Turner made a blurry painting called *Rain, Steam and Speed—The Great Western Railway*. Four years later, Karl Marx wrote *The Communist Manifesto*, declaring that the "constant revolutionizing of production, uninterrupted disturbance of all social conditions, everlasting uncertainty and agitation distinguish the bourgeois epoch from all earlier ones . . . All that is solid melts into air." He later expounded on this point, arguing that the continual speeding up of time was one of capitalism's fundamental dynamics, which he famously described as the "annihilation of space by time."

So complete was the nineteenth century's anni-hilation of space and time that time itself had to be reinvented. Before the railroads, each place kept its own time. When the sun was overhead, the time was noon. In England, Oxford time was five minutes behind Green-wich, and Leeds was a minute behind Oxford. Dover, in the east, was separated from Penzance by half an hour.[1] But in order to coordinate their schedules and prevent accidents, the railways had to create their own system of time, "railway time," to bring the discipline of a shared clock to the towns and stations along their routes. Rail-way time was the first step in a grand reinvention of time, an international effort to engineer a centralized, common time. "Time coordination," explains physicist and science historian Peter Galison, "was an affair for the individual school buildings, wiring their classroom clocks to the principal's office, but also an issue for cit-ies, train lines, and nations as they soldered alignment into their public clocks and often fought tooth and nail over how it should be done."[2]

The creation of absolute time entailed the subjuga-tion of localities, regions, and nations to a centralized tick of a clock at the Royal Observatory in south London, Greenwich Mean Time. Not everyone wanted to go along: in 1894, the French anarchist Martial Bourdin tried to blow up the Royal Observatory (his bomb exploded pre-maturely, killing him). But by the end of the century, the transformation was nearly complete. "Time ceased to be a phenomenon that linked humans to the cosmos,"

explains Rebecca Solnit, "and became one adminis-tered by technicians to link industrial activities to each other."[3] Time became something that was both absolute and malleable. The second hands at the Royal Observa-tory ticked uniformly, but time-bending technologies from railroads to telegraphs could be used to manage, leverage, destroy, and create time anew, particularly in the service of warfare or profit.[4]

But the industrialization of time had had a curious historical partner—namely, the discovery, and sub-sequent invasion, of deep time. In 1788, James Hutton published his *Theory of the Earth*. By carefully catalog-ing geologic strata, Hutton began to realize that the earth was far older than the six thousand years biblical scholars had calculated. "The world which we inhabit is composed of materials not of the earth which was the immediate predecessor of the present but of the earth which . . . had preceded the land that was above the sur-face of the sea while our present land was yet beneath the water of ocean." From the deep histories inscribed in rocks, he said, "we find no vestige of a beginning, no prospect of an end."[5] Hutton did for geology what Galileo did for astronomy. He is credited with the discovery of deep time: the timescale of tectonic plates and gradual erosion, the forces that slowly sculpt the earth's surface over millions of years, creating mountains, canyons, continents, and seas. Just as the industrial age discov-ered deep time, it began to actively shape it.

The Anthropocene is the informal term for the period

in which humans began to systematically shape the earth's surface on a planetary scale, and when political and economic forces began to have geologic consequences. In his 1994 article "On the Efficacy of Humans as Geomorphic Agents," earth scientist Roger Hooke set out to assess the impact of human activities such as agriculture and mining on the earth's surface. He found something remarkable: over the last hundred years or so, humans have moved more sediment than has been moved by classical geomorphic processes. Agricultural erosion, he estimated, moves about seventy gigatons of sediment annually, almost twice as much as meandering rivers (at forty Gt/y). Mining and highway construction move more earth than plate tectonics, glaciers, wave action, and aeolian (wind) processes combined.[6] The implication was profound. Shortly after the Industrial Revolution, human activities eclipsed natural earth processes as geomorphic agents.

Other earth scientists have expanded Hooke's work, suggesting the idea of an "anthropic force" and proposing new fields of "neogeomorphology" or "anthropogeomorphology." Geologist Peter Haff explains, "Anthropic modification of landscape is a new and unique phenomenon," whose effect on the earth's surface may be as significant as the emergence of vascular plants four hundred million years ago. Contemporary scientists wanting to understand earth processes, he argued, might do well to study economics, sociology, demographics, and other fields usually associated with the "soft" social sciences.[7] In the Anthropocene, the price of gold futures determines whether mountains will rise or fall, farm subsidies and commodity prices influence the rate of erosion, and a thin layer of radioactive fallout from the era of atmospheric nuclear testing has put a human signature on the Earth's geologic strata.

The Anthropocene is a period of temporal contradictions, a period in which Marx's space–time annihilation chafes against the deep time of the earth. The coal-fired plants and mass-produced automobiles of the industrial age have remade the sky in their own image, concentrating carbon dioxide in the air and eating holes in the ozone. It is an age in which ever-faster economic and political forces have ever more enduring consequences. The timescale of climate change, which unfolds over thousands of years, contradicts the timescale of human experience, which we measure in years or lifetimes, and the timescales of capital accumulation, measured in quarterly profits, daily receipts, or even millisecond fluctuations on a high-frequency trader's monitor. With nuclear power and nuclear weapons, the timescale of geopolitics and strategic military planning intersects the deep time of nuclear waste, which poisons its surroundings for countless thousands of years. A cup of fast-food coffee is meant to be sipped for a few minutes, but its Styrofoam takes more than a million years to biodegrade.

A carbon-saturated atmosphere, nuclear waste, and Styrofoam cups inhabit space at the expense of time. They are historic agents, producing their own futures. Nuclear waste insists that wherever it is will be poison for tens of thousands of years into the future. In this way, it freezes time in its own particular way, insisting that wherever it is, the future will be toxic. In the Anthropocene, communications and transportation are technologies ruthlessly deployed to annihilate space with accelerating time, but in human activities—from coal burning to fast-food coffee-cup manufacturing—we find an equally pitiless annihilation of the future: the annihilation of time by space. The fractures and folds in spacetime that form where the timescales of economics and politics collide with the deep time of the Earth are emblematic of the contemporary moment.

But what of culture? Structures like the Great Pyramid of Giza, Notre Dame of Paris, Stonehenge, and the Moai sculptures of Easter Island have far outlived the architects and workers who carved their stones. Cro-Magnon cave paintings and petroglyphs have lasted tens of thousands of years; Renaissance paintings have lasted hundreds. An albumen photograph from the late nineteenth century looks as good today as it did at the moment it was printed. But a chromogenic print from the 1970s has already faded, and a Zip drive from the late 1990s has become unreadable. Perhaps it is true that the more colorful and pictorial we make the world around us, the more ephemeral that excess of images becomes.

Over the last two centuries, humans have created 3.5 trillion still photographs, and we add hundreds of billions each year. In 2011, Facebook held more than 140 billion photographs, a repository ten thousand times larger than the Library of Congress.[8] On an average day, a person living in a city sees over five thousand images.[9] Pictures form the visual atmosphere of our daily lives. But our pictures are fleeting and elusive. In the far future, bits of hard drives may be fossilized in limestone, and discarded iPhones may find themselves encased in amber, hardened like nail polish, but the bits of humanity that these exquisitely crafted machines hold will be lost to time.

Earth's new moon, *EchoStar XVI*, embodies the Anthropocene contradiction between the hyperspeed of capital and the deep time of anthropogeomorphology, the torrential flow of twenty-first-century pictures and their utter ephemerality. While it whirls and hums with energy over its fifteen-year lifespan, *EchoStar XVI* will broadcast more than ten trillion pictures and video frames to earth-based televisions and computer screens.[10] All of these pictures will be as evanescent as the radio signals that carry them. But *EchoStar XVI* holds other pictures. A modest collection, to be sure, but one designed to last far longer than the oldest cave paintings. A collection designed to transcend the Anthropocene and to transcend deep time itself. A collection of pictures designed for the time of the cosmos. A collection of pictures that very well may be the last.

NOTES

1. Dan Falk, *In Search of Time* (New York: Thomas Dunn Books, 2008), 70–71; see also http://wwp.greenwichmeantime.com/info/railway.htm.

2. Peter Galison, *Einstein's Clocks, Poincaré's Maps* (New York: Norton, 2003), 40.

3. Rebecca Solnit, *River of Shadows: Eadweard Muybridge and the Technological Wild West* (New York: Penguin, 2004), 61.

4. For the space-time dialectics of capitalism, see David Harvey, *A Companion to Marx's Capital* (New York: Verso, 2010), 37; David Harvey, *Spaces of Global Capitalism* (London: Verso, 2006).

5. John McPhee, *Annals of the Former World* (New York: Farrar, Straus and Giroux, 1998), 79.

6. Roger LeB. Hooke, "On the Efficacy of Humans as Geomorphic Agents," *GSA Today* 4(9) (1994): 217, 224–225.

7. Peter Haff, "Neogeomorphology, Prediction, and the Anthropic Landscape," draft paper. Retrieved from http://www.duke.edu/~haff/geomorph_abs/neogeomorph%20paper/neogeomorphology.pdf.

8. Jonathan Good, "How Many Photos Have Ever Been Taken?" September 15, 2011. Retrieved from http://blog.1000memories.com/94-number-of-photos-ever-taken-digital-and-analog-in-shoebox.

9. Louise Story, "Anywhere the Eye Can See, It's Likely to See an Ad," *New York Times*, January 15, 2007. Retrieved from http://www.nytimes.com/2007/01/15/business/media/15everywhere.html.

10. This estimate assumes that *EchoStar XVI* broadcasts thirty frames per second on one thousand channels over fifteen years.

The Clarke Belt

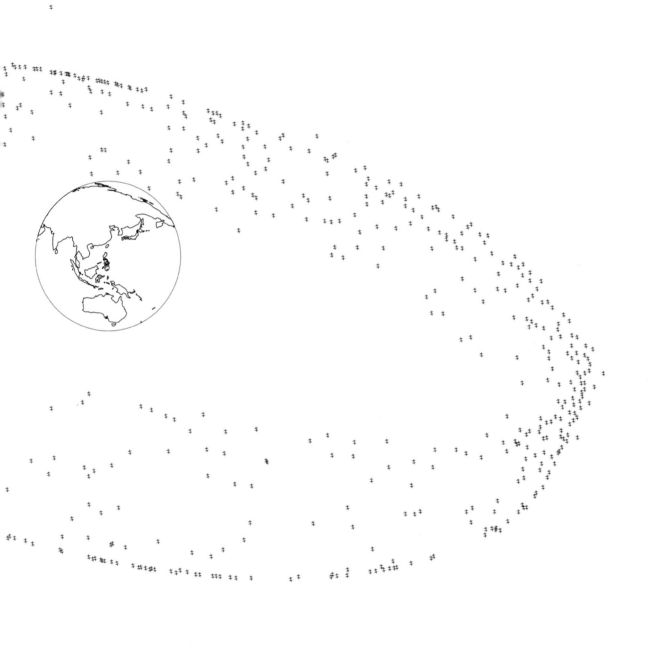

1 | ANCIENT ALIENS

On an upper floor of a nondescript high-rise in Toronto, I sat behind Ted Molczan's desk, peering over his shoulder at an Excel spreadsheet on his large LCD monitor. A scatter chart plotted the distribution of flash timings made by a satellite that he had spent years observing. Molczan, one of the world's leading amateur observers—whose particular specialty involves tracking secret spacecraft—strongly suspected that this particular object wasn't actually much of a satellite. Instead, he believed, it was a Mylar balloon deployed by the American military, a decoy to draw attention away from a highly classified "stealth" satellite code-named *MISTY*. The long flash timings indicated that the object was slowly rotating as it hurtled through space, nearly 3,000 kilometers (just over 1,800 miles) above Earth.

I'd travelled to Toronto to watch the night sky with Molczan, and he'd been a generous host. We watched the suspected decoy slowly amble across the sky, and we observed brighter and faster satellites designed for reconnaissance missions. Molczan showed me a castaway rocket body hurtling through the heavens and the faint flicker of an unknown object in a transfer orbit; he showed me how to generate precise observational data using little more than a stopwatch. But something else was also on my mind that evening.

Weeks earlier, in one of our sporadic phone calls, I'd asked Molczan if he had a good algorithm for determining satellite decay rates. When a satellite goes into orbit, I wanted to know, how long does it stay there? One thing I'd begun to understand about satellites, orbits, and celestial mechanics is that, in the words of theorist Jim Oberg, "space is quite literally 'unearthly.'" In his somewhat obscure but, in my opinion, important book, *Space Power Theory*, Oberg emphasizes that if you want to understand space, you need to understand that things in space play by different rules than things on earth. Understanding space, Oberg explains, means understanding that "much ordinary 'common sense' doesn't apply. One has to be cautious about making analogies with 'everyday life' . . . space is a physical frontier, it's also a mental one."[1]

From my conversations with Molczan, I'd come to understand something about what we might call the "geography of space"—or "orbitology," as Oberg terms it. Orbitology describes the topology of orbital space: the routes, passages, plateaus, parking slots, choke points, and gravitational hills and valleys produced through the interactions of celestial bodies.[2] In a sense, orbitology is the geography of rockets, satellites, and spacecraft. In my conversations with satellite observers like Molczan, I had begun to understand the unearthly nature of space's topology. But I'd come to suspect something else. Not only did "geography" work differently in space, but time itself was unearthly as well. And that's why I'd asked Molczan about how long things stay in orbit.

Ted Molczan is more than capable of elaborating on the finer nuances of celestial mechanics, but he didn't exactly answer my question. Instead of providing me with a cryptic algorithm that I was probably ill-equipped to understand anyway, Molczan suggested that I look into something called the *RAE Table of Earth Satellites*. Published by the U.K.'s Royal Aerospace Establishment and Defence Research Agency from 1957 until 1992, the *RAE Table* is an extensive catalog of man-made objects in Earth orbit. Like many other space catalogs, such as those published by the American Air Force and European Space Agency, the *RAE Table* records the names, altitudes, and orbital characteristics of thousands of satellites. But what's unique to the *RAE Table* is a number in its third column: how long a particular object is expected to stay in orbit.

By the time I got my hands on it, the *RAE Table* had been out of date for years, but that didn't matter. Molczan suggested that I simply look up the altitude of a satellite in one of the more contemporary catalogs and compare that orbit to something similar in the *RAE Table*. That comparison would give me a rough idea of how long any particular satellite would stay up in space.

For someone with mild obsessive-compulsive tendencies, perusing the spreadsheet-like *RAE Table* is absolutely fascinating. And the "lifetime" and "descent date" columns make it very clear that, indeed, time in space is deeply unearthly. The tables tell us, for example, that the discarded rocket body from a late Soviet-era satellite named *Cosmos 2133* (an "Oko"-class missile detection platform launched in 1991) was briefly in an orbit of 189 x 196 km (that is, an orbit whose low point, or perigee, is 189 km and whose high point, or apogee, is 196 km above Earth). At this low altitude, it took only two days for the higher reaches of earth's atmosphere to pull the spent rocket body back to the ground. In contrast, we learn that *Momo 1* (MOS-1), a Japanese marine observation satellite launched in 1987 to a 910 km-orbit, is predicted to linger in space until the year 2087. Small differences in altitude translate into hundreds of years in orbit. *Cosmos 1821*, a navigation satellite similar to the American GPS system, was also launched in 1987 into a slightly higher orbit of 963 x 1016 km. Although it's only a few kilometers higher than *Momo 1*, *Cosmos 1821* will remain in orbit until the year 2587—less than a hundred kilometers in altitude translates into five hundred years more time in orbit.

If we go up another few hundred kilometers, into the 1406 x 1467 km orbit of *Cosmos 1081* (a Soviet "Strela" military communication satellite launched in 1979), then the orbital time increases to seven thousand years. For a reality check, that number marks the amount of time between the rise of agriculture in the Nile valley and the present day. The kick-motor from an American Thor Delta launch in 1984 (*Star-37E PKM*, in a 558 x 49642 km orbit) will come down in about a hundred thousand years. That's three times longer than Aurignacian cave paintings, the oldest examples of figurative art on our planet. One hundred thousand years ago, Neanderthals roamed Europe, and the earliest recognizably modern humans were beginning to spread. And the only reason that kick-motor will fall back to earth so "quickly" is because its perigee is reasonably deep in the ionosphere, and the accumulated atmosphere drag slowly adds up.

And then there's the geosynchronous orbit, or GSO. This orbit was first theorized by the Russian aerospace visionary Konstantin Tsiolkovsky in the late nineteenth century, and the Slovenian engineer Herman Potočnik first calculated its distance in 1928. It's a special orbit, because a geosynchronous satellite orbits the earth at the same rate at which the earth rotates. Thus a GSO satellite effectively stays over one area on the earth's surface, and for that reason, GSO is the standard orbit for hundreds of telecommunications satellites broadcasting television, relaying phone calls, facilitating bank and credit card transactions, and piping music to the earth below.

The GSO is thin: satellites in GSO must remain within a narrow band only a few kilometers wide and high. The magic number where it all works is 35,786 km above mean sea level. This orbit is now home to hundreds of spacecraft; the GSO forms a ring of satellites around the earth, sometimes called the Clarke Belt in honor of science fiction writer Arthur C. Clarke (credited with first describing the GSO's potential as a platform for communications satellites in a 1945 article for *Wireless World* entitled "Extra-Terrestrial Relays: Can Rocket Stations Give Worldwide Radio Coverage?"). The communications satellites making up the Clarke Belt are like a man-made ring of Saturn forged from aluminum and silicon spacecraft hulls.

With the GSO, the *RAE Table* maxes out. It has two answers for the orbital lifetime of a spacecraft in GSO: "greater than a million years" and "indefinite." Because they experience virtually no atmospheric drag, these spacecraft will stay in orbit for unimaginably long periods of time.

All of this was in the back of my mind that evening in Ted Molczan's living room. After a lengthy discussion of the secret satellites that are Molczan's specialty—the imaging satellites in low-earth orbits with code names like Keyhole and Onyx—the conversation turned toward some of the GSO's secret denizens. The GSO isn't just home to much of the world's telecommunications infrastructure; it's also home to enormous eavesdropping satellites that take a special interest in the world's telecommunications. These spacecraft, with code names like Vortex, Magnum, and Mentor, lurk silently among the communications satellites in the Clarke Belt, eavesdropping on radio signals emanating from the earth below. Their giant umbrella-like antennae, designed to suck up even the faintest radio signals, are reported to be the size of football fields. Molczan said that these satellites, invisible even with binoculars and most telescopes on account of their great distance from earth, were the most secret satellites in Earth's orbit. I mentioned the great time scales I'd found in the *RAE Table*s for these spacecraft. "Yes," Ted concurred, "I think about them as artifacts."

Amateur satellite observers like Molczan have long understood that just beyond the reach of their binoculars lie some of human civilization's greatest engineering, and that these machines are effectively frozen in time. More than the cave paintings at Lascaux, the Pyramids of Giza, the Great Wall of China, the ancient city of Çatalhöyük, or the nuclear waste repository outside Carlsbad, New Mexico, the ring of abandoned satellites far above the earth's equator will house human civilizations' longest-lasting artifacts. In fact, there's really no comparison with any other human invention.

Derelict spacecraft in geosynchronous orbit

The hulls of spacecraft in the GSO do not measure their existence in seconds, hours, and years of human time, or even the thousands and millions of years of geologic time. Instead, they inhabit the cosmic time of stars and galaxies. Their ultimate fate may lie with the sun. About four billion years from now, the Sun will have burned through most of its hydrogen and will start powering itself with helium. When that happens, our star will swell to become a red giant swallowing the earth (and any lingering geosynchronous satellites). But four billion years is a long time from now. For a bit of perspective, four billion years is about sixteen times further into the future than the advent of the dinosaurs was in the past; it is four times longer than the history of complex multicellular organisms on earth. Four billion years is almost as far in the future as the formation of planet Earth is in the past. When Jim Oberg points out that space is "unearthly," he's right in more ways than he meant. Just as the topology of space is at odds with everyday human experience, the "time" of space is utterly foreign.

Placing a satellite into geosynchronous orbit means placing it into the deep and alien time of the cosmos itself.

What, if anything, does it mean that the spacecraft we build are undoubtedly humankind's longest-lasting material legacy? What does it mean that, in the near or far future, there will be no evidence of human civilization on the earth's surface, but our planet will remain perpetually encircled by a thin ring of long-dead spacecraft? Perhaps it means nothing. Or perhaps the idea of meaning itself breaks down in the vastness of time.

On the other hand, what would happen if one of our own probes found a graveyard of long-dead spacecraft in orbit around one of Saturn's moons? Surely it would mean something. What if we were to find a spacecraft from a different time—a spacecraft that contained a message or provided a glimpse into the culture that produced it?

These questions are the stuff of science fiction; they are themes in stories from H. P. Lovecraft's *At the Mountains of Madness* to Arthur C. Clarke's *Rendezvous with Rama* and Stanisław Lem's *His Master's Voice*. But they're also themes and questions on the outskirts of "real" science. Dozens of peer-reviewed articles are devoted to the topic in the academic literature. Notional fields such as "xenoarchaeology" and "exo-archaeology" propose applying methods from archaeology to alien artifacts that may be found in the future.[3]

In the early 1980s, technologist Robert Freitas and astronomer Francisco Valdes undertook a search for near-Earth extraterrestrial probes, coining the term "search for extraterrestrial artifacts" (SETA) to describe their efforts.[4] On November 6, 1991, astronomer and asteroid-hunter James Scotti found the first "candidate" object in a heliocentric orbit very similar to Earth's. Observations of the object, dubbed 1991 VG, showed that it was solid, exhibited a rapid variation in brightness, and was spinning rapidly. The object caught astronomer Duncan Steel's attention, and he began studying it. At first, Steel believed that 1991 VG was man-made, perhaps a leftover rocket body from a past space mission. But he quickly determined that no previous mission profiles seemed to fit. Next, he considered asteroids, but the similarity of 1991 VG's orbit to Earth's was unheard of. Furthermore, Steel showed that over long periods of time, 1991 VG's orbit would be disrupted by Earth itself. 1991 VG had to have entered its present orbit in the relatively recent past. Lacking an explanation for a terrestrial or natural origin for the object, Steel concluded, in a somewhat tongue-in-cheek article for the *Observatory*, that 1991 VG "is a candidate alien artifact" and that "the alternative explanations—that it was a peculiar asteroid, or a man-made body—are both estimated to be unlikely, but require further investigation."[5] The discovery of Earth Trojan asteroids (asteroids in stable orbits preceding Earth) in 2010, most notably an asteroid called 2010 TK7, seems to provide a solid explanation for the obscure object discovered in the early 1990s; 1991 VG is probably in the same family of asteroids as 2010 TK7. But at the moment, 1991 VG is too faint to observe. It won't be visible again until its next close encounter with Earth in 2017.

Talk of alien artifacts, interstellar probes parked in near-earth orbits, and dead spacecraft floating like ghost ships occupies a gray area between science fiction and hypothetical fields of social and physical science.

But that's true only from our particular temporal vantage point. We haven't yet found artifacts from an alien civilization in orbit around the earth, or on nearby planets and moons, but we are creating them. In the short term (ten to fifteen years or so), a communications satellite allows us to watch television stations, make phone calls, and the like. In the longer term (fifteen years to five billion years), it's a time capsule. It is a fragment of culture frozen in time, a future alien artifact.

In the future, we are the ancient aliens.

NOTES

1. Jim Oberg, *Space Power Theory* (Washington, DC: Government Printing Office 676-460, March 1999), 3.

2. Fraser MacDonald, "Anti-Astropolitik—Outer Space and the Orbit of Geography," in *Progress in Human Geography*, 31(5) (2007), 592–615.

3. Vicky A. Walsh, "The Case for Exo-Archaeology," in *Digging Holes in Popular Culture: Archaeology and Science Fiction*, Ed. Miles Russell (Oxford: Oxbow Books, 2002), 121–8.

4. Robert A. Freitas Jr. and Francisco Valdes, "The Search for Extraterrestrial Artifacts," *Acta Astronautica* 12(12) (1985): 1027–1034; Robert A. Freitas and Francisco Valdes, "A Search for Objects Near the Earth-Moon Lagrangian Points," *Icarus* 53 (1983): 453–457.

5. Duncan Steele, "SETA and 1991 VG," *The Observatory* 115 (1995), 78–83.

Sometime around 1910, a storm near the French town of Montignac uprooted a pine tree, revealing a deep cavity in the earth. Local farmers quickly filled it up with sticks to prevent their sheep from falling in. Thirty years later, the story goes, a local woman removed the sticks so that she could throw her dead donkey into the hole. She watched as the beast's body fell into the darkness. The cavity was much larger than she'd expected. Believing that she'd uncovered an old underground passage leading to a nearby chateau, the woman spread word of her discovery around town.

Then, one day in September 1940, four boys set out to explore the underground passageway. They didn't find any secret Middle Age architecture, but they did find something remarkable. In the flickering torchlight, the boys could see animal figures covering the cavern's walls. The cave held a tapestry of prehistoric art that had been hidden for seventeen thousand years. The boys were enraptured. "Our joy . . . was indescribable," one of them recounted. Another told interviewer Georges Bataille that they "felt like someone discovering a treasure, a casket of diamonds or a cascade of precious gems."[1] The boys had discovered the cave at Lascaux, now recognized as one of the world's greatest prehistoric troves of art.

For much of the time I've spent researching the images and ideas for *The Last Pictures*, a print showing one of Lascaux's most enigmatic paintings has hung above my desk. I've looked at it every day and thought about all the different things it could mean, if indeed it means anything at all.

THE PIT

This particular painting lies in one of Lascaux's deepest recesses, past the Hall of Bulls, through the Passageway, and beyond the Apse, in a secluded corner that archaeologists and prehistorians have come to call the Shaft or the Pit. The image lies in utter darkness at the bottom of a fifteen-foot drop, where the air is thick with carbon dioxide. The painting is so strange that it's difficult to describe without unintentionally interpreting it. From right to left, there is a bison, its head lowered and tail raised. A circular shape hangs under the bison's abdomen. A line is drawn through (or over) the bison's rear section. In front of the bison is an ithyphallic stick–figure humanoid with an inordinately small and forward-pointing head. The bird-like head of the man had led some scholars to dub him the bird-man; others have called him the dead-man because he seems to be falling backward in front of the bison. Just below the bird-man or dead-man is a bird with a vertical line below it, drawn in the same stick-like style as the man. To the left is what looks like an incomplete painting of a rhinoceros facing away from the bird-man, bird-stick, and bison. Under the rhinoceros's tail are six dots, evenly arranged in two rows.

This Pit scene is unique. Prehistorian Norbert Aujoulat, who directed research at Lascaux for a decade, explains that the greater cave system contains 915 ani-

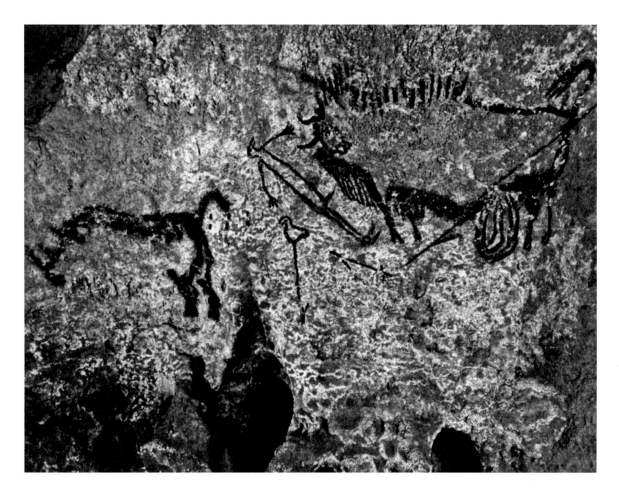

mal figures, 434 signs, and 613 depictions of indetermi-
nate figures. In total, there are 1,963 representations.
Of these nearly two thousand images—most of them
without any obvious referent—the bird-man is the only
humanoid.[2]

This Pit scene has several common interpretations.
Abbé Henri Breuil reads the scene as a hunt gone wrong:
a disemboweled bison and half-dead man are left in
the wake of a rampaging rhinoceros, which is "moving
away peacefully after having destroyed all that annoyed
it."[3] Other scholars, notably Mircea Eliade and David
Lewis-Williams, see shamanistic motifs in the painting:

"Sex is sometimes associated with shamanistic travel,"
explains Lewis-Williams of the bird-man's erection.[4]
Jean Clottes adds to the shamanic interpretation of the
Pit scene: "Death—that of the bird-man and the bison—
is obviously a prominent theme of this mysterious panel,
and the bird a significant motif. Traditional cultures
often conceived of both death and shamanic trance as
the flight of the soul from the body, which birds often
symbolized."[5] Clottes adds that altered states of con-

sciousness might be involved, pointing out that the Pit is "a place prone to extremely high rates of carbon dioxide, which may cause hallucinations and acute discomfort."[6]

My own interpretation of this painting changes from day to day, hour to hour, and is based on nothing more than letting my mind wander while staring at the print on my wall. My thoughts vacillate. Sometimes I see an image that's profoundly unknowable. A painting from tens of thousands of years ago surely lies beyond some horizon of intelligibility. Such a painting, separated from the world that gave rise to it, can only elude interpretation. On those days, I question the act of trying to make sense of the painting. Does the concept of *sense* apply to such an image? Does the image have meaning, or does this image live outside the world of things that mean anything at all?

On those days, I see something profoundly unknowable, something utterly alien. Indeed, this is the opinion of Norbert Aujoulat, the person who probably knows Lascaux better than any other scholar: "[W]e have succeeded in identifying and decoding only a fraction of its message . . . [the painting] remains all the more difficult to grasp in that some of its numerous aspects lie on the threshold with the irrational."[7]

If the painting lies beyond concepts such as meaning, representation, and intelligibility, then the prehistorians who've written so much about this image have done so in vain. Whatever this painting may have been is irretrievably lost, and nothing connects the present to the deep past from which this painting came.

At other times, however, the universe seems a little less profound, things seem more connected, and the image seems to speak. On those days, I can imagine that the painting captures something about the deeply intense relationship between a nearly naked human,

armed with little more than a few pointy sticks, and a bison, or humans' relationship with nature in general. Perhaps this painting represents an erotic encounter with death, signified by the falling man and a disemboweled bison. But maybe the bison is not disemboweled at all, and the shape toward its rear legs is actually the bison's penis and testicles or an engorged vulva. In this case, perhaps, the image is a lament over a fissure between humans and animals, and at the same time a celebration of the continuities between them.

But perhaps this is not a hunt scene at all. Perhaps it is, in fact, a murder confession. I can imagine a Cro-Magnon artist looking at the world around her, a world where humans had begun to systematically exterminate the megafauna with which they shared the land. Before long, mastodons, aurochs, cave lions, cave bears, wooly rhinoceroses, and most of the European bison would be extinct. Perhaps the painting shows a man enraptured and stimulated by the megadeath he has brought to the surrounding landscape. Perhaps the artist knew that killing the rest of the world was ultimately a suicidal proposition. Maybe this painting is hidden in the darkest recesses of a cavern because it is a secret admission of bloodlust and guilt, like the scrawled handwriting of a serial killer on a note to the police, half pleading, half taunting: "Stop me before I kill more; I cannot control myself."[8]

Or not. Sometimes I imagine something far simpler—that the person who made it was simply insane. Or a big joke: the bird-man is drawn in a stick-like style because it's the work of a prehistoric vandal; the sexually excited bird-man is like a penis drawn with a sharpie on a subway advertisement.

Finally, I've thought that this enigmatic painting may not have been intended for its own time at all, but perhaps was intended for the far future. Maybe this painting was actually made for us. That would explain why it lies so deep in the recesses of a primordial cavern. Maybe the cave's depths are the prehistoric equivalent of outer space. Indeed, some prehistorians have hypothesized that the majority of unintelligible marks at Lascaux, the vast collections of dots and lines, are a coded guide to Cro-Magnon stars.[9]

The Pit painting may have been born out of futility and frustration. When the ancient artist or cave-cosmonaut set out to paint an image for the future, she may have realized the ridiculousness of such a proposition, the insanity of what she'd set out to do. And so she decided to paint an image that was insane. She could have imagined a time in the far future where her painting would be seen as a key to understanding "early humanity," and, realizing the absurdity of that, painted a picture to impart the following message to us: early humanity, whatever that is, is absurd. The painting's meaning might be precisely its own meaninglessness.

Maybe that is what we're supposed to learn from the Pit at Lascaux.

THE GREAT RING OF DEAD MACHINES
Communications satellites never cease to fill me with wonder and awe. I find them deeply strange, unsettling, and *untimely*. While they are powered up and active, these spacecraft route signals and broadcast media at the speed of light, connecting noncontiguous places across the globe to one another, creating new geographies. Images and sounds rush through their titanic hulls, which face down toward Earth from their position high above the planet. But when these electronic megaliths power down and die or use the last of their fuel to boost themselves into a graveyard orbit just above the

Clarke Belt, they become the stuff of a future real-life science fiction drama. They join a ring of dead spacecraft ruins destined to remain in Earth orbit until Earth is no more. Like other spacecraft, they will far outlast anything else humans have created.

This project, *The Last Pictures*, was inspired by the idea that we should take communications satellites seriously as the cultural and material ruins of the late twentieth and early twenty-first centuries. When Creative Time asked me to develop an idea for an artwork having to do with space, I proposed developing a cultural artifact and attaching it to a future ghost spacecraft. The artifact's message would be a story about what happened to the people who built the great ring of dead machines around Earth. The message would not be a grand representation of humanity. It would not be a portrait of life on earth. Instead, the message would be a riff on an observation made by the British historian Arnold Toynbee: "Civilizations die from suicide, not by murder."

I thought about the early Rapa Nui people of Easter Island, who erected iconic statues (Moai) that look endlessly up toward the sky. To create the ropes, timber sleds, ladders, and levers needed to move the Moai, the Rapa Nui felled every last tree on their island home. The deforestation decimated the local ecosystem, rendering Easter Island nearly uninhabitable. Maybe the early Rapa Nui understood perfectly well that their statues were a means of collective suicide and went ahead and kept building them anyway. Until they cut down the last tree to make the last rope to erect the last monolith.[10]

I printed out a text and stuck it on my studio wall: "How is it that we knew exactly how we were going to kill ourselves, and went ahead with it all anyway?"

The entire premise of the project was, of course, absurd. The idea that someone in the future might actually find the Artifact was close to nil. The notion that the message could actually mean anything at all seemed ridiculous. The message could only be a failure. For me, that wasn't the question. The question was whether it could be an *interesting* failure. The probability of the artifact having an audience in the far future was almost nil, but the probability of people on Earth thinking about it here and now was guaranteed.

From the moment I began thinking about dead satellites and suicidal civilizations, the ghost machines in Earth orbit became an allegory for what happened to the people responsible for them. An allegory for the recent history; perhaps even an allegory for modernity itself. Like the suicide statues at Easter Island, the dead satellite embodied a cruel paradox: perhaps the interactions of production, technologies, and forms of knowledge that allowed us to explore the heavens also enabled us to destroy our own island Earth. The science that enabled us to understand the insides of stars was put to practical use in the inner workings of a hydrogen bomb; the rockets that took us to space, we designed primarily to deliver thermonuclear Armageddon. Of course, many people envision that humans will indeed commit thermonuclear suicide or exhaust Earth's resources. For some, theis represents "progress." In this view, it is our collective destiny to colonize other planets. The dream of colonizing space is a farce, but it fuels the ideology that humans are separate from the earth and that destroying our own habitat is a problem we can outrun.[11] Perhaps the satellites we've established in the heavens are our own version of the Easter Islanders' Moai statues.

I decided that the artifact I was planning could only be a grand gesture about the failure of grand gestures. The message should be a collection of images—a slideshow for eternity.

CONVERSATIONS

The project I envisioned was too big for one person. On a practical level, I had no experience with aerospace engineering or materials science. On a philosophical level, I knew that a great number of people in a broad range of fields had already spent a lot of time thinking about the questions I was asking. I also knew that critical thinkers would be skeptical of the project. The notion of creating something timeless or universal is the stuff ideologies are made of.

One person who immediately came to mind was the biologist Ignacio Chapela at UC Berkeley. Chapela became a public figure in late 2001 when he and graduate student David Quist published an article in *Nature* showing how genetically modified corn imported from the United States was infecting traditional strands of Mexican corn. The biotechnology industry launched a massive smear campaign against him. *Nature* retracted the article under industry pressure, and UC Berkeley, which had recently received a $25 million grant from biotech giant Novartis, denied Chapela tenure despite the support of his department (his tenure case was overturned on appeal).

Chapela was uneasy about my project: he didn't think that such a grand gesture could inspire much critical thought. Big gestures, like tinkering with life to create genetically modified Frankenfoods, were very much a part of the problem.

Others agreed. Author Mike Davis warned me that he couldn't imagine a project "that doesn't risk reproducing the solipsism of Carl Sagan's Rosetta Stone for aliens" before suggesting what became one of my favorite ideas: that artifact be a "a few cc's of mothers' tears" sealed in a vial of Trinitite (the glass created at Trinity, when desert sand melted under the world's first nuclear explosion).

UC San Diego cognitive scientist Rafael Núñez was deeply skeptical of any gesture toward universal communication. A specialist in how humans do mathematics, Núñez spent the better part of a year teaching me the math I was supposed to have learned in high school, then showing me how the entirety of mathematics emerges from a series of spatial metaphors that arise from our bodies' interaction with the surrounding world. Our ideas of math and numbers are an eminently human set of inventions. He enjoyed making fun of science fiction authors and astronomers who imagine that prime numbers or transcendental numbers such as pi might provide a basis for communication with extraterrestrials.[12]

Despite a general unease with the project among some of the people I approached, others were willing to entertain a set of sustained conversations around the question of "how the humans committed suicide." But, as was to be expected, the questions multiplied: Was this a project about global warming and ecological collapse? Or about technology gone out of control? Was it a conversation about the Cold War or Big Science, or was it a question about the nation-state? Was it a conversation about the Enlightenment? About capitalism? About modernity? Or about the old myth of the naked people who ate from the forbidden tree of knowledge?

I began holding weekly seminars in the Creative Time offices, where research assistants Emily Parsons-Lord, Katie Detwiler, Max Symuleski, Laura Grieg, Anya Ventura, and I looked at thousands of images and held long conversations about everything from cybernetics to messages in bottles, from medieval bestiaries to various kinds of mathematical infinities. Often the discussion topic was more abstract: what the heck were we doing, exactly? On several occasions we had guests come speak to our group. Sundar Sarukkai of Manipal University led an outstanding seminar on the history of Indian mathematics, showing how provincial the metaphysical assumptions underlying Greek-descended geometry are. Moreover, Sarukkai argued that our common assumption that curiosity is a value-free trait is, in fact, a deeply ideological and relatively recent notion. (For Sarukkai, the idea that scientific *discovery* is a good thing in and of itself is a rather dangerous proposition. If scientific truths are *discovered*, then the *discoverer* bears no responsibility for them. If, however, science is *invented*, then the people doing the inventing have a great deal of responsibility for the implications of their work. If, for example, nuclear physics is discovered, then Szilard, Fermi, Oppenheimer, Teller, and their compatriots are absolved of any responsibility for the advent of nuclear weapons. If, however, nuclear science is invented, then they bear a great deal of responsibility.[13])

As we sifted through thousands of pictures, sometimes going to a great deal of effort to locate a particular image or document, we came to realize something else. Biologist Susan Oyama put it well in a conversation about ecological collapse: the things that most threaten us are those for which there are no images. What does a picture of global warming look like? (A terrified polar bear on a piece of melting ice?) What does rampant resource depletion look like? (A clear-cut rainforest?) What sort of picture signifies ecological destruction? (An aerial image of an oil spill?). What is a photograph of economic inequality? (Portraits juxtaposing the lives of rich and poor?) What does a picture of capitalism look like? (A factory spewing filth into the sky? A day trader in front of a computer terminal?)

My thoughts continually came back to the Lascaux bird-man. I thought of the Pit scene as not just a painting from prehistory, but a painting about the power and limitations of images. The scene seems to imply a narrative, but there is none. It creates an impression but explains nothing. Moreover, what's true of the Pit is true of all images: they can't explain or narrate much of anything at all. Instead, they ask us to see what we're predisposed to see. In this way, all images are like cave paintings.

OUTSIDE OF TIME

In weekly seminars, our group discussed all sorts of ancient and contemporary artifacts: eighth-century Islamic astrolabes, fifteenth-century alchemical manuals, present-day livestock cloning kits, Renaissance cryptography handbooks, tomes of medieval magic, late twentieth-century eschatological maps, nineteenth-century proto-calculators, and twenty-first-century source code for weaponized computer viruses. We came back to a few objects so often that they became major touchstones for our own process. A source of almost endless interest was the work of a handful of scientists, artists, and journalists, who in the 1970s did their best to design a series of messages for aliens. A few years later, some of these same people applied their theories to a series of postapocalyptic warning signs in New Mexico.

In the 1970s, space probes were things of discovery and wonder, the stuff of a better future. When a series of missions put spacecraft on trajectories that would ultimately take them out of the solar system, NASA commissioned a series of visual greeting cards to be attached to their probes.

For the 1972 and 1973 *Pioneer* spacecraft, a small group—including Carl Sagan, Linda Salzman-Sagan, and astronomer Frank Drake, a luminary in the search for extraterrestrial intelligence (SETI)—developed a postcard for any aliens that might find the spacecraft on its grand voyage through interstellar space. Salzman-Sagan drew a naked couple, with the man raising one hand, standing in front of an outline of the spacecraft. Below them, a drawing of the solar system attempted to show *Pioneer* leaving the third planet from the sun. To explain where the probe came from, Frank Drake developed an ingenious galactic map based on newly discovered celestial objects called pulsars. Pulsars are collapsed stars that spin rapidly, giving off intense, directed radiation. They are often compared to cosmic lighthouses. Drake figured that if he provided the unique time signatures of various pulsars, extraterrestrial scientists could use them to triangulate the location of Earth.

In 1974, Frank Drake created another message intended for extraterrestrials to celebrate the opening of the refurbished Arecibo radio telescope in Puerto Rico. This message took the form of a powerful radio transmission directed at the star cluster M13. The message was a series of 1,679 bits, which are decoded by arranging them into 23 columns and 73 rows. Drake reasoned that because 1,679 is a semi-prime number, divisible only by prime numbers 23 and 73, the correct arrangement of rows and columns should be self-evident. Properly decoded, the message forms a graphic reminiscent of an old Atari video game. The image is meant to convey information about the chemical makeup of DNA, along with the double helix DNA structure, a human figure, the solar system, and a picture of the Arecibo radio telescope.

But the grandest "postcards for aliens" are undoubtedly the Golden Records that were placed onboard the two *Voyager* spacecraft in 1977. Developed by a committee led by Carl Sagan, these gold-plated copper phonograph records are intended to explain something

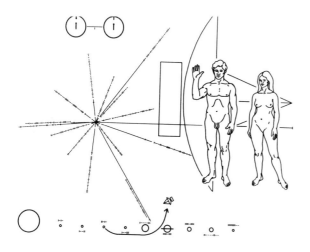

about human life and culture to any extraterrestrials who might stumble across the spacecraft in the distant future. Timothy Ferris, one of the record's coauthors, explained that the LP is "intended to preserve something of human culture beyond what an intelligent extraterrestrial, encountering the craft at some far-distant time and place, might infer from the spacecraft itself."[14]

One side of the record holds a collection of 116 photographs encoded as a video signal. It is a smorgasbord of snapshots: the Great Wall of China, the Taj Mahal, a woman in a supermarket eating a grape, dolphins jumping, diagrams of human anatomy, rush hour traffic, a Chinese dinner party, a sunset. The record's other side is a collection of world music: a Navajo night chant; Western classical music from Bach, Beethoven, Mozart, and Stravinsky; Javanese gamelan; Senegalese percussion; Peruvian pan pipes; Chuck Berry. There are greetings in dozens of languages—"Hello from the children of planet Earth" (English); "Friends of space, how are you all? Have you eaten yet? Come visit us if you have time" (Amoy); "Good day to all" (Spanish)—and a selection of sound effects from Earth: "fire and speech," "tame dog," "kiss," "Morse code," "crickets, frogs," and "mud pots."[15]

Today the Pioneer Plaques and Voyager Golden Records continue hurtling through space at nearly 35,000 miles an hour relative to Earth. Their dose of "It's a Small World" benevolence remains ready to explain something about life on earth to any extraterrestrial contacts.

With the end of the 1970s and the onset of the 1980s, however, some of the people responsible for the iconic record reconvened for a rather darker thought experiment. At the Waste Isolation Pilot Plant (WIPP) near Carlsbad, New Mexico, semioticians, anthropolo-

Top: The Pioneer Plaque

Bottom: The Arecibo message (decoded)

"Spikes Bursting Through Grid," concept by Michael Brill, art by Safdar Abidi

gists, scientists, and science fiction writers have spent decades developing a set of signs intended to last, and to be legible, for the next ten thousand years. Their message is simple: *Stay away; this site holds invisible death.* The Waste Isolation Pilot Plant is an underground chamber filled with nuclear waste.

The warning sign project began in 1981, when the U.S. Department of Energy and Bechtel Corporation convened a series of working groups to "determine whether reasonable means exist (or could be developed) to reduce the likelihood of future humans unintentionally intruding on radioactive waste isolation systems."[16] The various groups included Frank Drake, who developed the Arecibo message and the pulsar map for *Pioneer* and *Voyager,* and the artist Jon Lomberg, who oversaw much of the visual material for the Golden Record. Carl Sagan was asked to join, but refused to participate on moral grounds.[17]

The two teams came up with a number of proposals. Architect-planner Michael Brill imagined large-scale "menacing earthworks," composed of "shapes that hurt the body and shapes that communicate danger." Brill imagined giant spikes emerging from the desert floor, a "landscape of thorns," tangles of spikes and steel. Another group, "Marker Group, Team A" proposed a less grandiose approach: just write warning texts in multiple languages. Team A pointed out that "since literacy first developed 6,000 years ago, it has not ceased to exist" and that "scholarship capable of translating the messages on the markers will continue to exist somewhere in the world during the time period being considered."[18] Another Team A suggestion was to design a star chart showing how the danger from nuclear waste contamination diminishes as the celestial pole moves away from Polaris toward Vega over the next twenty-six thousand years.[19]

But one of the most surprising proposals for how to relay a warning to the future came from the person with the deepest knowledge of how symbols and signs work. Semiotician Thomas Sebeok recommended against using images or symbols for timeless warning signs. In the words of communications professor Peter C. Van Wyck, Sebeok "saw clearly the futility of merely launching a sign into the future."[20] But this didn't discourage him from tackling the problem at hand. He merely took another approach, swapping semiotics for theater and performance. To warn future people about the dangers of nuclear waste, Sebeok wanted to convene an "atomic priesthood" to act as the site's caretakers. Members of this priesthood

 DANGER
POISONOUS RADIOACTIVE ☢ WASTE BURIED HERE
DO NOT DIG OR DRILL HERE BEFORE A.D. 12,000

Proposed warning sign, Waste Isolation Pilot Plant, New Mexico

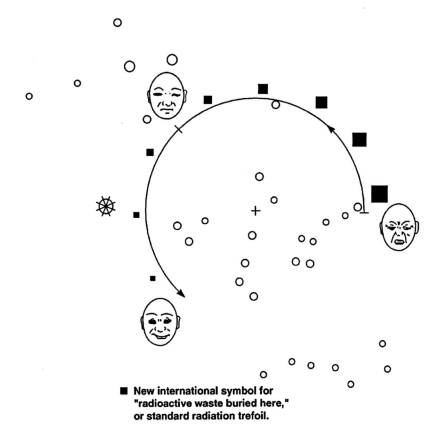

■ New international symbol for "radioactive waste buried here," or standard radiation trefoil.

would invent myths and stories as they saw fit to secure the nuclear waste not only in physical space, but also in the imaginations and stories of future people. Knowledge about the nuclear site would be passed from generation to generation among the initiates of a mystery cult charged with protecting the site's secret.[21]

STAR CHARTS

Back in New York City—1,800 miles from the New Mexico nuclear waste site and about 11 billion miles from the *Voyager* spacecraft—curator João Ribas from the

Star map intended to show passage of time and decrease in radiation over thousands of years

List Visual Arts Center at the Massachusetts Institute of Technology (MIT) heard about our project and invited me to MIT to work on the project. One of the technical problems I was trying to solve was that of archivability: what materials and techniques exist, if any, to produce an "ultra-archival" medium for storing images? Project manager Alexis Lowry had been researching materials and fabrication; at one point we were exploring the idea of writing photographs into the crystalline structure of a diamond. Then we found out it would cost more than a hundred thousand dollars to fabricate. We needed something more affordable, but had diamond-like properties, such as an exceptionally stable atomic structure, the ability to survive the stress of a rocket launch, drastic temperature fluctuations in space, and the constant radiation bombardment it would experience in orbit. Professor Brian Wardle of MIT's Aeronautics and Astronautics Department volunteered to help.

During our first phone call, I asked Wardle about making something that would be archival for five billion years. "That's actually a new question," he responded in an upbeat voice. Wardle enlisted Professor Karl Bergreen of the Materials Science and Engineering Department, an expert on nano-fabrication techniques. It didn't take them long to settle on a solution: we'd create a finely-etched silicon disc using a nano-fabrication process typically used to build microscopic circuits. We could modify that process to imprint photographs on a small silicon wafer. It would be extremely archival—and spaceworthy. I was also pleased that the images would be visible without any complicated decoding techniques. They'd be visible to the unaided human eye and would come to life under a magnifying glass or microscope. Our object would be like an old microfiche, but one that could theoretically last for an eternity. The only downside was that working at a human scale (as opposed to a truly nano-scale) would limit the number of images we could write. I thought that was a good tradeoff—fewer images, but a far greater chance that someone would be able to actually see them.

As fabrication and delivery deadlines started looming, I had to decide what the disc cover should look like. In our weekly discussions, my group and I had decided that the cover should do two things. First, it should be obvious that the disc cover isn't a part of the host spacecraft—it should call attention to itself, and it should look out of place. Second, it should look like a piece of treasure. It should look valuable.

Of all the decisions I made in researching and developing the project, the one that surprised me the most was what the final cover etching would look like. I always thought the cover would be something deliberately surrealistic, a nonsensical image or pattern. At one point I thought the cover should be an image of a tall, goat-headed man towering over a startled child. But as the deadline for the final designs got closer and closer, I started to have a dramatic change of mind. At another, I thought the cover should bear a simple inscription: "Please do not disturb me. Let me stay here so that I may witness the end of time."

A few months earlier, I'd met astronomer Joel Weisberg at a series of events at Carleton College in Minnesota. Joel was from the same generation of astronomers who'd worked on the '70s-era extraterrestrial postcards. He'd been good friends with the late Val Boriakoff, the man eating a tuna sandwich in my favorite image from the *Voyager* collection: a bizarre photograph meant as a "demonstration of licking, eating, and drinking." I knew that Joel was a pulsar specialist, and I knew that Frank Drake had used pulsars as the basis for his *Pioneer* and *Voyager* maps. Joel and I began working on a time-map design for the cover plate (see page 174 for a full description).

I'd never imagined a time map being part of the project. I always thought that the chance of anyone ever being able to decode the weird drawings onboard *Pioneer* or the proposed nuclear warning signs in New Mexico were very close to zero. I saw those charts and signs as the quixotic scratchings of people who believed in irrational things. The people who made these signs may or may not have believed in God, but they believed, or desperately wanted to believe, in something *universal,* whether it be physics, chemistry, astronomy, geometry, or transcendental numbers like pi.[22] We can't even talk to dolphins or dogs, I reasoned, how are we supposed find some sort of universal baseline from which to begin a conversation with aliens?

But I knew that could be wrong. After all, humans talk to dolphins and dogs every day. We don't talk to dogs about pi, but we certainly talk to them (incidentally, we don't spend a lot of time talking to other humans about pi, either, and when we do it's not clear that we make much sense). I started to feel haunted, not by the fact that our artifact would be unintelligible, but on the contrary, that someone might actually find it and be able to make some sense of it. A phrase popped into my head from John Knowles' novel *A Separate Peace*, which I'd read in grade school: "Always say some prayers at night because it might turn out that there is a God." What was there to lose by making a sincere effort to inscribe some temporal coordinates into the artifact? If the disc cover was deliberately designed as nonsense, then the chance of someone ever decoding it was zero because there would be nothing to decode. True, even if the scratching did have an underlying meaning, the chances of it ever meaning anything to anyone were close to zero. Nonetheless, the remote possibility of someone finding and being able to make sense of it kept nagging at me.

I'd framed the question of what to put on the cover as one about semiotics and the production of meaning, but maybe there was something else at stake with the artifact cover. I began to feel that if I made something deliberately nonsensical, I would be abrogating some sort of potential responsibility toward the future. Maybe, I thought, the question of what to do with the cover wasn't a matter of meaning, but a matter of ethics.

One thing that always puzzled me about the astronomers and scientists who put together the 1970s messages to extraterrestrials was their utter sincerity. They really did try their best to develop a language that an extraterrestrial might be able to understand. They truly wanted to share images and information about humans with other potentially "intelligent" life forms in the universe. I always thought their sincerity was at best naïve and at worst potentially even colonialist. But I started to wonder whether I had become too cynical.

I sheepishly skyped Rafael Núñez at UC San Diego, the cognitive scientist who'd generously spent nearly a year explaining to me why we couldn't use mathematics as the basis for any universal language. I told him that I was making a star map for the artifact cover and was planning to use a version of Drake's notation, as a nod to the 1970s messages. Núñez was visibly annoyed. "But what if we're wrong?" I said. I suggested that it might be best to "pray at night," even if we didn't believe in God. If we make a star map, and no one ever finds it, then we've lost nothing. If we make a star map, and someone does find it but can't interpret it, then we also haven't lost anything. But if we're wrong and they *can* interpret it, the star chart might make them very happy. It might truly be a treasure.

With reluctance, Núñez started playing along with the idea, making it very clear that he didn't agree at all but was simply going along with my thought experiment, "praying at night." He helped me decide on the "calibration" shapes for the artifact cover: two geometric shapes meant to "teach" our notation to someone who might find the disc. We settled on a figure of a right triangle whose sides are 1, meaning its hypotenuse is the square root of two. This would be a reference to a well-known (at least on Earth) ancient Babylonian math tablet called YBC 7289. The other geometric figure shows the proportion of the area of a circle to a surrounding square—a figure chosen because, unlike transcendental numbers such as pi and e, it has absolutely zero metaphysical baggage.

And so the artifact's gold cover is a star map. As such, it may echo the hopeful pulsar map of the 1970s space probes. Or the darker star map of the 1980s, designed to show how many eons must pass and how far the stars must move before the twentieth century's nuclear waste ceases to poison the surrounding landscape. But, more precisely, the artifact's cover is not a star map at all. It is an ensemble of basic shapes, dots and dashes, lines and assorted squiggles etched onto a gold-plated piece of aluminum. It becomes a star chart only if someone chooses to see it that way. In this sense, the cover etching simply recapitulates the inscrutable scratchings, paint marks, lines, and dots that make up the majority of images on the walls of Lascaux. The line through the buffalo, the six dots under the rhinoceros's tail, and the barbed line under the bird-man's feet. After all, these marks may, in fact, be star charts.

NOTES

1. Georges Bataille, *The Cradle of Humanity* (New York: Zone Books, 2005), 95–97.

2. Norbert Aujoulat, *Lascaux: Movement Space and Time* (New York: Abrams, 2005), 257.

3. Abbé Breuil, *Four Hundred Centuries of Cave Art*, trans. Mary Boyle (Dordogne: Centre D'Etudes et de Documentation Prehistoriques – Montignac, 1952), 135–136.

4. David Lewis-Williams, *The Mind in the Cave: Consciousness and the Origins of Art* (London: Thames and Hudson, 2002), 265.

5. Jean Clottes, *Cave Art* (London: Phaidon, 2008), 120.

6. Ibid.

7. Aujoulat, *Lascaux*, 254.

8. This idea echoes Bataille, 2005.

9. David Whitehouse, "Ice Age Star Map Discovered," *BBC News*, August 9, 2000. Retrieved from http://news.bbc.co.uk/2/hi/871930.stm

10. Jared Diamond, *Collapse* (New York: Penguin, 2005), 79–120.

11. This statement echoes Hannah Arendt's sentiments in her essay "The Conquest of Space and the Stature of Man." In Hannah Arendt, *Between Past and Future* (New York: Penguin, 1993).

12. See George Lakoff and Rafael Núñez, *Where Mathematics Comes From* (New York: Basic Books, 2000) and with Reuben Hersh, *What Is Mathematics, Really?* (Oxford: Oxford University Press, 1997).

13. Sundar Sarukkai, "Science and the Ethics of Curiosity," *Current Science* 97(6) (2009): 756–767.

14. Timothy Ferris, "The Mix Tape of the Gods," *New York Times*, September 5, 2007. Retrieved from http://www.nytimes.com/2007/09/05/opinion/05ferris.html?_r=3&oref=slogin.

15. Carl Sagan et al., *Murmurs of Earth: The Voyager Interstellar Record* (New York: Random House, 1978), 154–157.

16. Human Interference Task Force, *Reducing the Likelihood of Future Human Activities That Could Affect Geologic High-Level Waste Repositories* (Columbus, OH: Office of Nuclear Waste Isolation, May 1984), BMI/ONWI-537.

17. Peter C. Van Wyck, *Signs of Danger* (Minneapolis: University of Minnesota Press, 2005), 59.

18. Kathleen M. Trauth et al., *Expert Judgement on Markers to Deter Inadvertent Human Intrusion into the Waste Isolation Pilot Plant* (Albuquerque: Sandia National Laboratories, November 1993), SAND92-1382, 3-2.

19. Trauth et. al., F118.

20. Van Wyck, 47.

21. Thomas Sebeok, "Pandora's Box: How and Why to Communicate 10,000 Years into the Future," in *On Signs*, ed. M. Blonsky (Baltimore: John Hopkins University Press, 1985).

22. Pi is an especially poignant image here—in Carl Sagan's book *Contact* (New York: Simon and Schuster, 1985), an advanced extraterrestrial civilization explains that a vast and transcendental mystery is buried somewhere in pi's infinite digits.

"BELONGING": HUMAN/ARCHIVE/WORLD

BY KATIE DETWILER

From our earliest discussions of what images we might include in the Artifact, we were faced with a question of belonging. This collection, or archive, would be defined not only by what it contained but also by what was excluded; the division of what belongs inside the archive and what does not would be constitutive of the archive itself.

At first our group set a provisional boundary for the archive: no images of the human form would be included; all human figures would be removed or obscured. What would we mean to do by excluding the human form? Different people had different answers to this question. We never wanted or intended the collection of images to be a grand representation of humanity. Removing the human altogether seemed to be a way to avoid this kind of sweeping gesture. The Artifact would be something that would exist well beyond the tenure of humans on earth, so excluding them would imply this and be a way to think about this scenario. Finally, the human tends to be privileged as the central feature of the world. Could the collection, the message, be more unstable and generative without the anchor of the human as the locus and sole purveyor of meaning? What if some other form entirely—why not a machine, or the empty desert?—could be the image through which some Other would know us?

A friend offered a metaphor for what we were doing by excluding humans from this object: "You're building the Flying Dutchman." The Flying Dutchman: a ship at sea, a human construction, yet with no human hands to man it; a ghost ship. Here, the human would be a trace, no longer in control of steering the course of meaning. The Artifact would also echo other things humans have set in motion; things that humans may have helped create, but that now seem to function without our notice, understanding, or direction.

But we started to grow suspicious of our idea of making the human ghostly. The idea that we can make ourselves invisible is a very human fantasy. It's the idea that we can suppress the relationship between ourselves and the things we produce, a fantasy of invisible hands that organize and curate without themselves being implicated. The politics of this could backfire. Moreover, if we didn't include images of humans, we'd be implying that humans are somehow able to step outside of the world and ourselves and see things from a detached vantage point. Isn't it disingenuous to pretend that the human is something we can step outside of and look at as if it were an object?

This turned out to be a complicated question. Some would say that this dual subject-object position actually characterizes the human fairly well: We belong to a category that we ourselves have named. We belong in the world according to organizing structures that we ourselves have invented. Depending on how one thinks about this doubled position (and within our own group we disagreed about it), this is a kind of human specialness or a kind of specialness that is also a form of alienation; an exceptionalism that makes us uncomfortable in our own skin and makes our relationship to the world complicated because we see ourselves both as part of it and above it, slightly estranged from it. Shouldn't we (and how could we?) try to represent this double position of humans as archive-makers who can include or exclude themselves, subjects and objects; this position of belonging and exceptionalism?

Alongside these questions, we were encountering another kind of problem as well. If we had started by wanting to avoid a kind of narcissism about humans as the stable center of the world, our attempt to exclude humans pushed us into a position of having to be very confident about when a human form appears. This seemed to be just another way of shoring up and affirming the human as something objectifiable, well-defined, and stable. It also proved difficult. We tend to see "the human" reflected in all sorts of nonhuman things. Does an image of an apparition of a face on the surface of Mars contain a human form? Is a disembodied human brain cradled in latex-gloved human hands a human form? We found that we could answer these questions in a number of ways. If we had thought that we might literally cut humans out of images of bodies, machines, and nature, not only did we start to feel that this was another way of separating the human out and casting it out from the world, but we also grew more and more uncertain about where to even make such cuts.

We eventually had to reframe our idea of excluding the human, realizing that neither human/not-human nor exclusion/inclusion are simple dichotomies.

At our final meeting, we ended up in a discussion of one image that made many of us most uncomfortable. Not necessarily an image of a human form, it's an image of Captain America: a comic book superhero and the alter ego of a frail human. Grotesquely muscular, nearly bursting the seams of his garish costume, he is a mutant. This super-human avenger had an interesting career: at one point purportedly assassinated, later an intelligence agent, born and revived from suspended animation. How like the career of the human.

By the end of the project, we decided to include images of humans in the Artifact. The human belongs to the archive as one artifact among others, another liminal object. It does not stand outside of the collection or outside of composite scenes of bodies, machines, and nature. But, in the end, it seemed equally honest to include something of ourselves as we also are: narcissistic about our humanness and our exceptionalism, archive-makers also exceptionally proud of our reflexivity about our own boundaries.

3 | ONE HUNDRED PICTURES

The image collection for the *EchoStar XVI* Artifact was selected though a process of interviews and conversations, archival research, formal considerations, and aesthetic sensibility. Many of the images' themes emerged from interviews with physical scientists, social scientists, artists, activists, and philosophers. In some cases, these conversations coalesced into a single image, but in most cases they helped set the project's overall tone and served as thematic guides to the process. In addition, a research group made up of Anya Ventura, Emily Parsons–Lord, Katie Detwiler, Laura Grieg, and Max Symuleski spent nearly five months combing through libraries, databases, and archives collecting images and ideas and discussing them at weekly seminars. The image collection is meant not only as a small archive, but also as a montage: a silent film for the darkness of space and the depths of time.

מוזיאון ישראל, ירושלים
the israel museum, jerusalem
متحف اسرائيل، اورشليم القدس

Paul Klee, 1879-1940
Angelus Novus, 1920
Oil transfer and watercolor on paper
The Israel Museum Collection
Gift of John and Paul Herring, Jo Carole
and Ronald Lauder, Fania and Gershom Scholem
B 87.994

ירושלים 91710, ת"ד 71117, טל. 02-708811, פקסימיליה 02-631833
91710 jerusalem, p.o.b 71117, tel. 02-708811, fax. 02-631833

220 EAST CHICAGO AVENUE • CHICAGO, ILLINOIS • 60611.2604
Museum of Contemporary Art

Negotiating Rapture: The Power of Art to Transform Lives
MCA, Chicago: June 21-October 20, 1996

X.96.34/Cat.34
Paul Klee
New Angel (Angelus Novus), 1920

12-1/2 x 9-1/2 in. (31.8 x 24.2 cm)
The Israel Museum

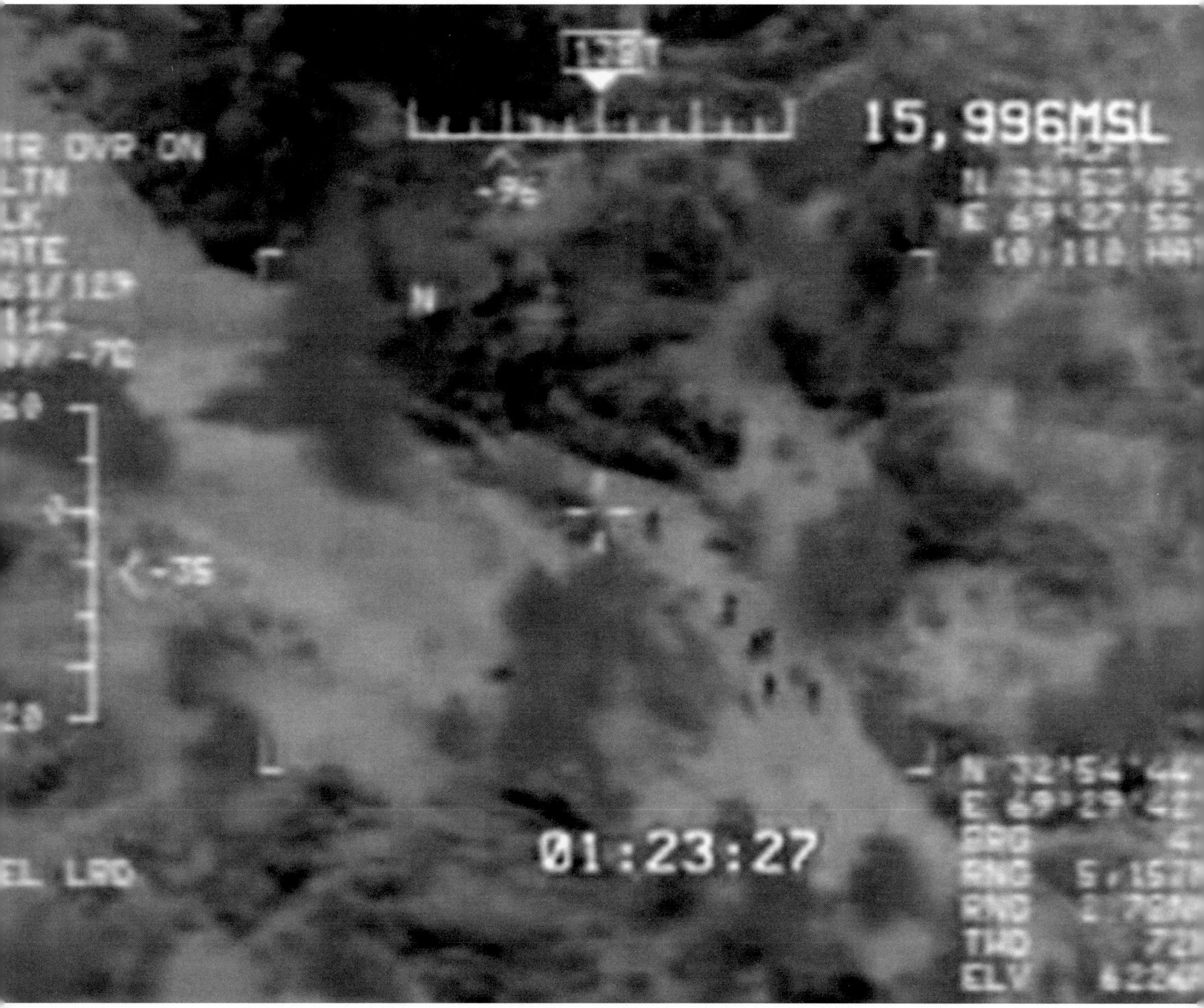
15,996MSL
01:23:27
ELV

West of House
You are standing in an open field west of a white house, with a boarded front
door.
There is a small mailbox here.

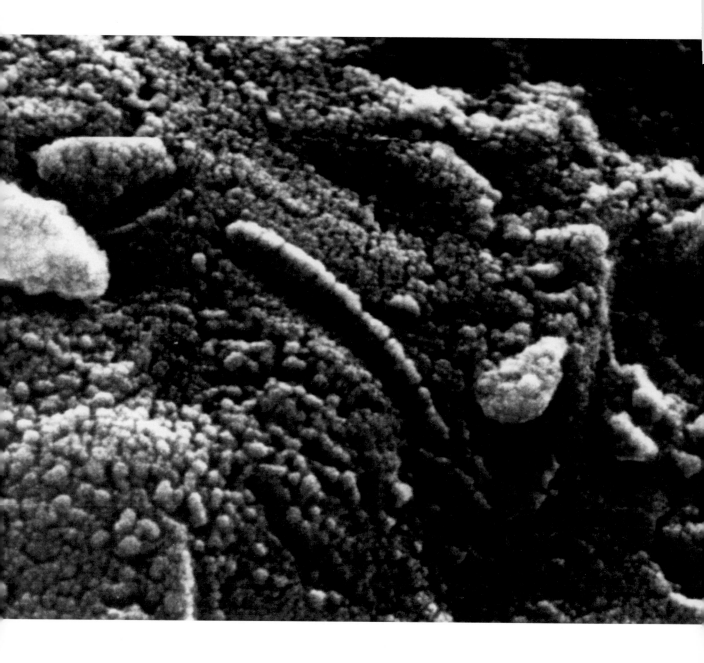

25kV   1m

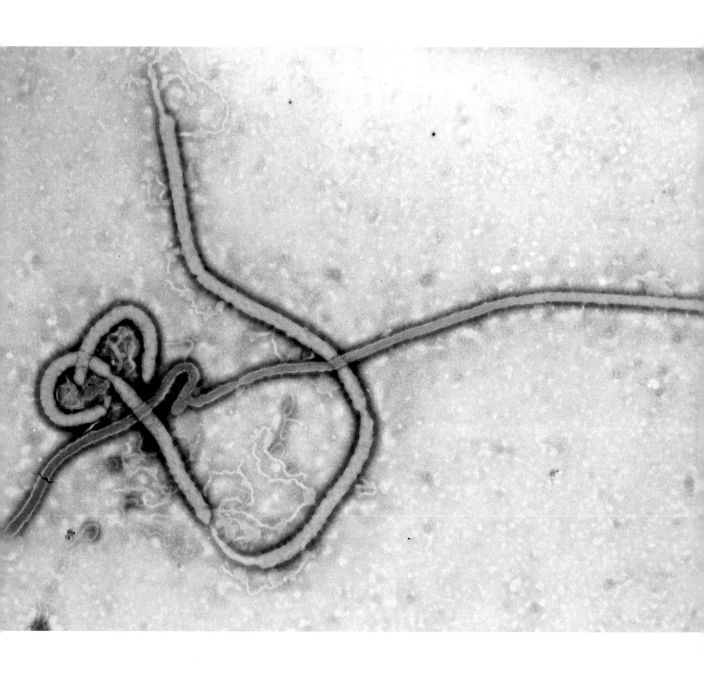

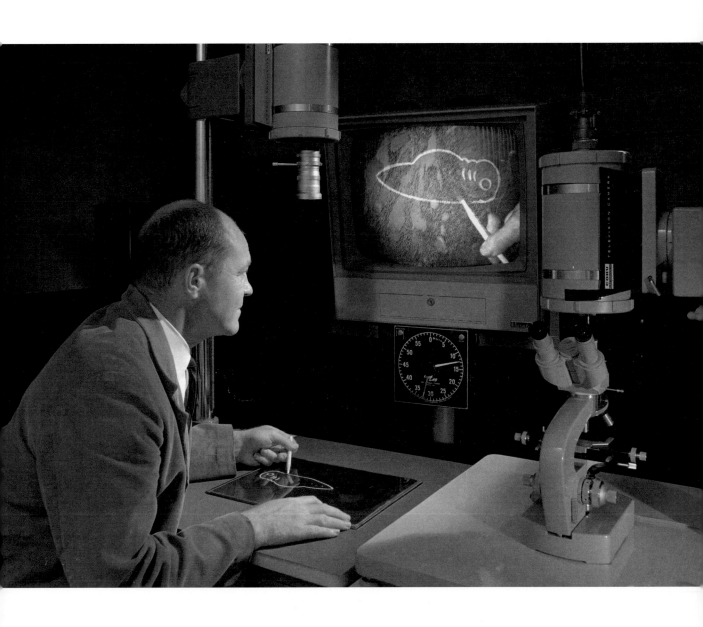

AMERICAN, BRITISH, FRENCH & GERMAN MASKS   465

9194

0800   antan started   { 1.2700   9.037 847 025
1000   "   stopped - antan ✓   9.037 846 795 correct
        13" (032) MP - MC   ~~1.9821647000~~
                                2.~~130476415~~ (-3) 4.615925059 (-2)
            (033)   PRO 2   2.130476415
                comct   2.130676415
        Relays 6-2 in 033 failed special speed test
        In Relay   "   10,000 test.
                Relays changed
~100 Started Cosine Tape (Sine check)
1525 Started Mult+ Adder Test.

1545   Relay #70 Panel F
                                                                (moth) in relay.

        First actual case of bug being found.
~~1545~~1630 antangent started.
1700 closed down.

Ledlano: very urgently.
Ledlik : reddish.
Ledlinot : mordant.
Ledom : palace; castle; mansion.
Ledomal: lord of the castle.
Ledoman : inhabitant of the castle.
Ledön : to redden; to color red.
Ledotik : scrupulous.
Ledotiko: scrupulously.
Ledotöf: scrupulousness.
Ledüfik : of stony hardness.
Ledüfugik : obdurate.
Ledük: grand-duke.
Ledükatäd : stock in trade; stock.
Ledul: perseverance; steadiness.
Ledulik : persevering; persistent.
Ledulön : to persevere; to endure; to hold out.
Ledunik: industrious.
Ledunöf: industry.
Ledunön : to do eagerly; to be in-
LEF : yeast.   [tent on.
Lefad : string; line; cord; twine.
Lefad fananuga : fish-line; fishing-line.
Lefadön : to tie with string ; to put string around.
Lefal : sudden fall; tumble; plunge.
Lefälokön : to precipitate one's self; to hurl one's self.
Lefalön : to fall; to be precipitated.
Lefälön : to precipitate; to push over.   [grand-sire (paternal).
Lefatel : great grand-father; great

Lefid : meal; food; repast.
Lefidön : to partake of a meal; to sit at table; to give a repast.
Lefikulik : very difficult.
Lefiled : conflagration.
Lefined : middle finger.
Leflen : intimate friend ; bosom friend; confidant.
Leflidel : Good Friday.
Leflum : current; stream.
Leflüm : electric current.   [run.
Leflumön : to stream; to flow; to
Leföa : anew ; afresh ; from the beginning.
Lefoetik : ravenous; savage.
Lefog: cloudiness; cloud.
Lefogik : cloudy.
Lefomam : finishing; perfecting.
Lefomön : to finish; to perfect.
Lefon : bubbling water ; bubble ; eddy; bubbling spring; foun-
Lefop : idiot ; lunatic.   [tain.
Lefopavut : raving madness; mania.   [cal.
Lefopavutik : raving mad; mania-
Lefopik : idiotic; demented; insane.
Lefopöp : lunatic asylum ; mad-house.   [insanity; dementia.
Lefopug : madness ; distraction ;
Lefot : forest of large trees.
Lefoviko: quick as lightning; with lightning rapidity.
Lefüd : east; orient.
Lefüdänön : to orientalize.

Least: luün; lunün; nemodikünos.
Least : nemodüno.
Least, at: leluüno.
Least of all : lelüno; leluostüno.
Least, in the : lunüno.
Leather: skit.
Leather : skitik.
Leather dealer: skitan.
Leather manufacturer : skital.
Leathery; skitlik.
Leave: däl; dismüt.
Leave, to: lüvön ; geletön ; la-iletön; fegelütön.
Leave, to grant: dismütön.
Leaved : bledemik.

Leaving: lüvam.
Leaves, to strip off: debledön.
Lechery : lenepuedug.
Lector : lilädal.
Lecture, to : bililädön.
Lecture room : tidöp.
Lecturer : dokal; plivadokal.
Leech : gib.
Lees : glusiadot.
Left : nedet.
Left : nedetik ; pelüvöl.
Left, at the : nedetiko.
Left, from the : nedetoa.
Left, on the : nedeto.
Left, on the — of : nedetü.

Lefüdan : eastern man; inhabitant of the east; oriental.
Lefüdik : eastern; oriental.
Leful : perfection; perfectness; accomplishment. [consummation.
Lefulam : finishing ; completion ;
Lefulamik : feasible; practicable.
Lefulamöf : feasibility ; practicability.
Lefulel : perfecter; improver.
Lefulik : perfect; accomplished.
Lefuliko : perfectly.
Lefulnik : complete; full; entire.
Lefulniko : completely.
Lefulno : abundantly.
Lefulnöf : completeness.
Lefulnön : to execute; to accomplish; to complete.
Lefulnot : completion.
Lefulo : perfectly.
Lefulön : to finish; to complete; to perfect; to consummate.
Lefulot : finishing; consummation.
Lefulsam · perfection ; improvement.
Lefulsel : perfecter; improver.
Lefulsön : to perfect; to improve.
Lefüved : eastward.
Lefüvedo : eastwardly.
LEG : genuineness ; authenticity ;
Legad : park.   [legitimacy.
Legäl : ecstasy; transport; rapture.
Legälam : enchanting; ecstasy.
Legälik : rapturous; transporting ; ecstatic.
Legälöl : enchanting; charming.
Legälölo : enchantingly; charmingly.
Legälön : to enrapture; to transport; to enchant; to charm.

LEGAT : embassy; mission.
Legatef : legation; embassy (persons).
Legatel : embassador; legate.
Legatem : embassies; missions.
Legatöp : office of legation ; embassy.
Legeil : summit; top; highest point.
Legeilik : elevated; high; lofty.
Legeilön : to top; to culminate.
Leget : conception.
Legidik : august; lofty; sublime.
Legik : genuine ; authentic ; legit-
Legion : legion.   [imate.
Legionan : legionary.
Legiv(am) : donation; presenting.
Legivel : donor; giver.
Legivön : to donate; to give.
Legivot : drink-money ; fee ; tip ; gift; gratuity (as to servants).
Leglav : groats; grits; pearl-barley.
Legleipön : to assail; to attack; to assault.   [mous.
Legletik : colossal; gigantic; enormously; immensely.
Legleto : enormously; immensely.
Leglif : grief; sorrow; affliction.
Leglifik : sorrowful; afflicted. [fret.
Leglifön : to grieve; to sorrow; to
Legödelo : in the early morning.
Legodö! Great God !
Legolöp : corridor.
Legudik : very good ; quite good ; excellent.   [harper.
Lehapal : skilful harpist; skilful
Lehät : helmet.
Lehel : ringlet; curl; lock.
Lehelakapil : little curly head.
Lehitüp : mid-summer.
Leif : even if.
Leiflanik : equilateral.

Left, to the : nedetoi.
Leg : lög; goap.
Legacy : gelet; gelütadil.
Legal : gitlik; lavogik; lonik.
Legalization : lonikam.
Legalize, to : lonikön.
Legate : legatel.
Legation : legatef; legatöp..
Legend : lusag; lekonil.

Legendary : lusagik.
Legible : lilädik.
Legibility : lilädof.
Legion : legion.
Legionary : legionan.
Legislate, to : lonön.
Legitimacy : gitlöf; leg.
Legitimate : gitlik; legik.
Legumes : veäduk.

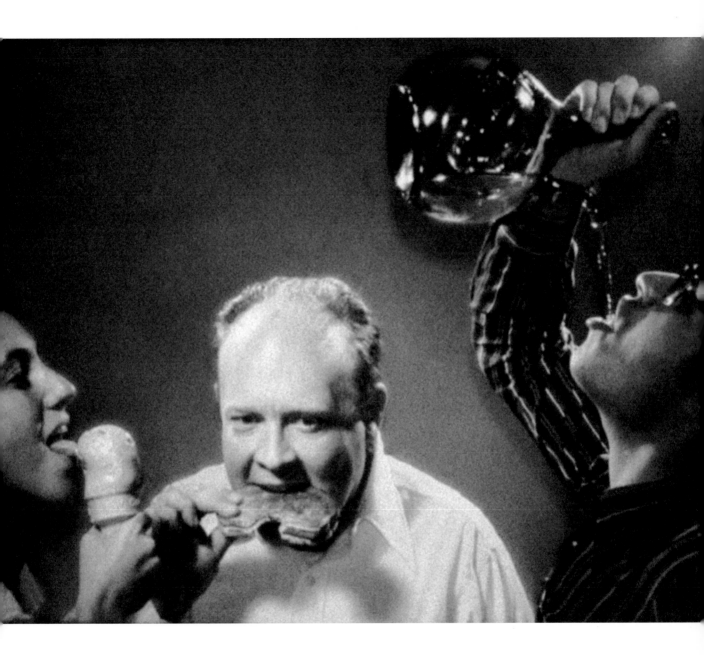

Besides the common sort of *Swans*, there is a wild kind, called *Hooper*, having the *wind-pipe* going down to the bottom of the *breast-bone*, and then reversed upwards in the figure of the Letter *s*.

Besides the common *Goose*, there are several sorts of *wild ones*, whereof one is *black* from the *breast* to the middle of the *belly*, called *Brant Goose*, *Bernicla*, or *Brenta*.

To the *Widgeon-kind* may be reduced that other *fowl*, about the same bigness, the two middle *feathers* of whose *train* do extend to a great length, called *Sea-Pheasant*, *Anas cauda acuta*.

To the *Teal-kind* should be reduced that other *fowl*, of the like shape and bigness, but being *white* where the other is *green*, called *Gargane*.

To the *Gull-kind*, doth belong that other *Bird*, of a long slender *bill* bending upwards, called *Avosetta recurvi rostra*.

Of Beasts.

δ. V. BEASTS, may be distinguished by their several shapes, properties, uses, food, their tameness or wildness, &c. into such as are either
Viviparous; producing living young.

{ WHOLE FOOTED, the *soles* of whose *feet* are undivided, being used chiefly for *carriage*. I.
CLOVEN FOOTED. II. *Pag*.157.
Clawed, or *multifidous*; the end of whose *feet* is branched out into *toes*; whether
NOT RAPACIOUS. III. *158*.
RAPACIOUS; living upon the prey of other *Animals*; having generally *six short pointed* incisores, or *cutting teeth*, and two long *fangs* to hold their prey; whether
CAT-KIND; having a *roundish head*. IV. *159*.
DOG-KIND; whose *heads* are more *oblong*. V. *160*
OVIPAROUS; breeding *Eggs*. VI. *161*

I. WHOLE FOOTED BEASTS.

Equus.

Asinus. Mulus.

*Lev.*11.4.26.

Camelus.

Elephas.

I. WHOLE FOOTED BEASTS, may be distinguished into such as
Solid hard hoofs; considerable for (are either of
Swiftness and *comeliness*; being used for riding.
  1. HORSE, *Mare, Gelding, Nag, Palfrey, Steed, Courser, Gennet, Stallion, Colt, Fole, Filly, Neigh, Groom, Ostler.*
Slowness and *strength* in *bearing burdens*; having *long ears*; either the more *simple* kind: or that *mungrel* generation begotten on a
  2. ASSE, *Bray.* (*Mare.*
  3. MULE.
Softer feet; having some resemblance to the
Cloven footed-kind; by reason of the upper part of the *hoof* being divided, being *ruminant*, having a *long slender neck*, with one or two *bunches* on the back.
  3. CAMEL, *Dromedary.*
Multifidous kind; having little *prominencies* at the end of the *feet*, representing *toes*, being of the *greatest magnitude* amongst all other *beasts*, used for the carriage and draught of great weights, and more particularly esteemed for the *tuks*.
  4. ELEPHANT, *Ivory.* II. CLOVEN

II. CLOVEN FOOTED BEASTS, may be distributed into such as *[II. CLOVEN FOOTED BEASTS.]*
Horned and Ruminant; having *two horns.* (are
Hollow; not branched nor deciduous, being common both to the *males* and *females*, useful to *men* both living and dead; whether the
Bigger; being useful both by their *labour* and *flesh*;
  1. KINE, *Bull, Cow, Ox, Calf, Heifer, Bullock, Steer, Beef, Veal, Runt, bellow, low, Heard, Cowheard.* *[Bos.]*
Lesser; being useful either in respect of the *Fleece and Flesh*: or *Hair and Flesh.*
  2. SHEEP, *Ram, Ewe, Lamb, Weather, Mutton, Bleat, Fold, Flock;* *[Ovis.]*
    GOAT, *Kid.* (*Shepheard.* *[Caper.]*
Solid; branched, deciduous, being proper only to the *males*; whether the
Bigger kind; || either that of the highest *stature*, having *horns* without *brow-antlers*, of a *short stemm*, and then spreading out into *breadth*, branched at the edges: or that of a *lower stature*, having round, long, branched *horns.*
  3. SELKE, *[Alche.]*
    STAGG, *Hart, Hind, Red Deer, Venison.* *[Cervus.]*
Middle kind; whose *horns* become broad towards the ends; || either that of *lesser horns*, not used for labour: or that which hath the *largest horns* in proportion to that *body*, of any other *Deer*, with a double branched *brow-antler*, being in the *Northern Countries* used for the *drawing of Sleds.*
  4. BUCK, *Doe, Fawn, Pricket, Sorel, Sore, Fallow Deer, Venison.* *[Dama.]*
    REIN-DEER, *Tarandu.* *[Rangifer.]*
Least kind; having a *short, round, branched horn.*
  5. ROE-BUCK, *Roe.* *[Capreolus.]*
Horned but not ruminant; having but one *horn*, placed on the *nose*, being a *beast* of great bigness, covered with a kind of *Armature*, and counted untamable.
  6. RHINOCEROT. *[Rhinoceros.]*
Ruminant but not horned; being useful to *men* only, when living, for carriage of burdens, having the *longest neck* of any other *Animal* (if there be really any such *Beast.*)
  7. CAMELOPARD, *Giraffa.* *[Camelopardus.]*
Neither horned nor ruminant; useful only when dead, for its flesh.
  8. HOG, *Swine, Bore, Sow, Pig, Porket, Barrow, Shoot, Pork, Bacon, Brawn. Grunt.* *[Porcus.]*

Amongst those that belong to the *Bovinum genus*, there are several sorts described by *Authors* distinguished by their having either

A Beard;		*Urus.*
A Bunch on the back;	stiled	*Bisons.*
Horns reflected about the ears.		*Bonasus.*
Broad, flat, rugged horns;		*Buffalus.*

Besides the more common kinds of *Sheep*, there are others mentioned by *Authors*, and described to have

Streight wreathed horns.	called	*Ovis strepsiceros.*
Great thick tails.		*Broad tailed Sheep.*

Amongst

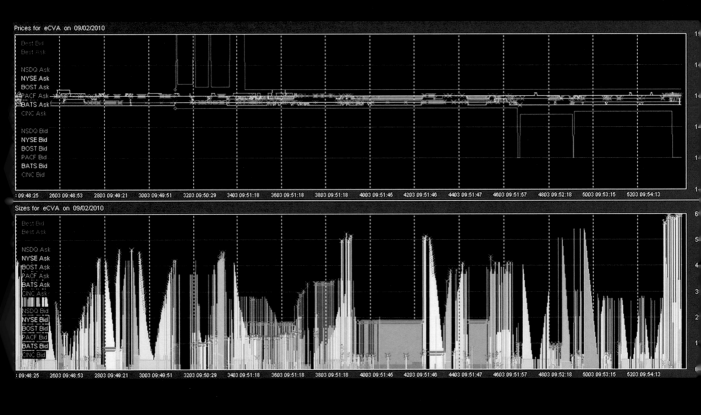

qui avait été donné à Palerme et dont le
était tout récent lorsqu'il habitait cette

positions des nombreux baudets qui, en Sicile,
vaient à effectuer tous les transports et il en ch

Fig. 1. — Un piano de chats. (D'après une gravure du dix-septième siècle.)

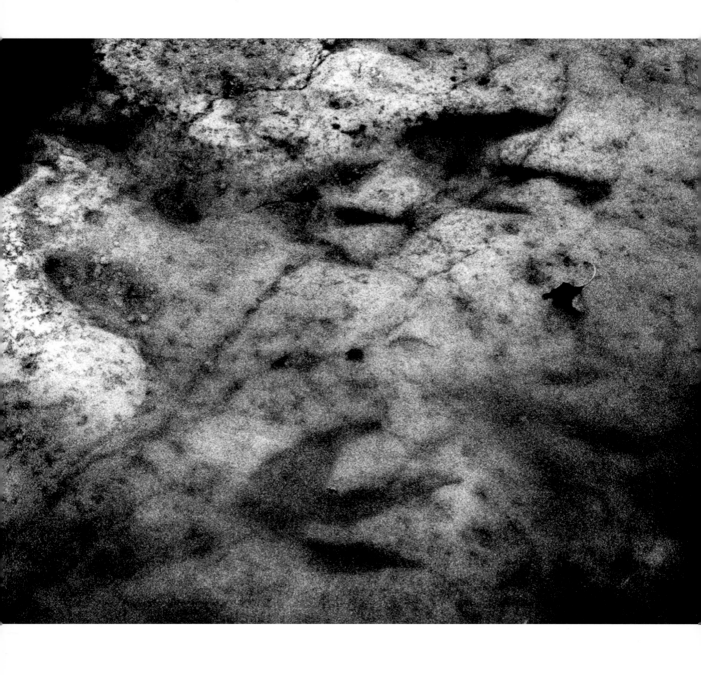

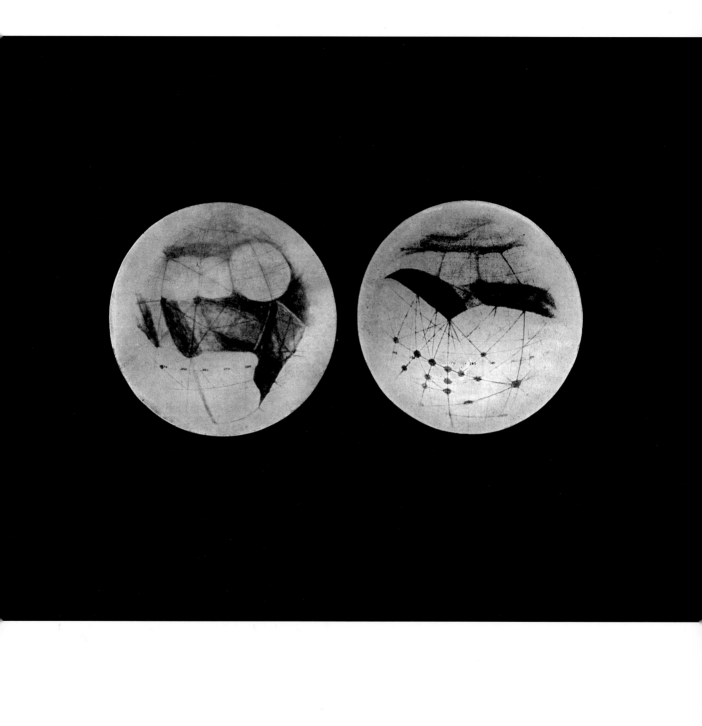

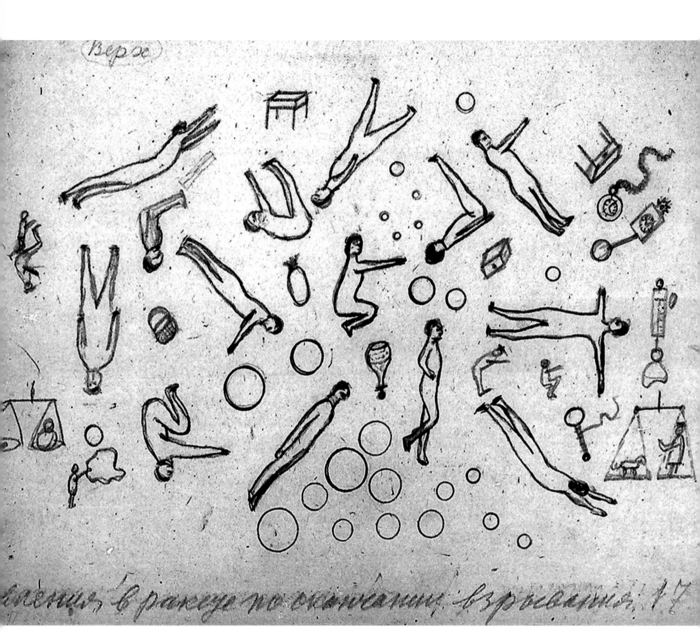

CHANNEL NUMBER (TWO DIGITS, WRITTEN VERTICALLY)
0000000000111111111122222222223333333333444444444455
1234567890123456789012345678901234567890123456789012

R T	RT ASCEN. 1950.0 HH MM SS	DECLIN. (1950.0) DD MM	2ND LO FREQ. (MHZ.)	GLCTIC LAT. (DEG.)	GLCTIC LONG. (DEG.)	EASTERN STD TIME HH MM SS	OBJECT
	19 05 22	- 27 02	120-162	-15.39	10.13	22 04 10	
	19 05 37	- 27 02	120-163	-15.44	10.16	22 04 25	
	19 05 49	- 27 02	120-163	-15.48	10.18	22 04 37	
	19 06 01	- 27 02	120-164	-15.52	10.19	22 04 49	
	19 06 25	- 27 02	120-164	-15.56	10.21	22 05 01	
	19 06 37	- 27 02	120-165	-15.60	10.24	22 05 13	
	19 06 49	- 27 02	120-165	-15.64	10.27	22 05 37	
	19 07 01	- 27 02	120-166	-15.72	10.28	22 05 49	
	19 07 13	- 27 02	120-166	-15.76	10.30	22 06 01	
	19 07 25	- 27 02	120-166	-15.81	10.32	22 06 13	
	19 07 37	- 27 02	120-167	-15.85	10.34	22 06 25	
	19 07 49	- 27 02	120-167	-15.89	10.36	22 06 37	
	19 08 01	- 27 02	120-168	-15.93	10.37	22 06 49	
	19 08 25	- 27 02	120-168	-15.97	10.39	22 07 01	
	19 08 37	- 27 02	120-169	-16.01	10.41	22 07 13	
	19 08 49	- 27 02	120-169	-16.05	10.43	22 07 25	
	19 09 01	- 27 02	120-169	-16.09	10.45	22 07 49	
	19 09 13	- 27 02	120-170	-16.18	10.46	22 08 01	
	19 09 25	- 27 02	120-170	-16.20	10.48	22 08 13	
	19 09 37	- 27 02	120-170	-16.30	10.50	22 08 25	
	19 09 49	- 27 02	120-171	-16.30	10.52	22 08 37	
	19 10 01	- 27 02	120-171	-16.34	10.54	22 08 49	
	19 10 13	- 27 02	120-172	-16.38	10.55	22 09 01	
	19 10 25	- 27 02	120-172	-16.42	10.57	22 09 13	
	19 10 37	- 27 02	120-173	-16.46	10.59	22 09 25	
	19 10 49	- 27 02	120-173	-16.50	10.61	22 09 37	
	19 11 01	- 27 02	120-173	-16.54	10.63	22 09 49	
	19 11 13	- 27 02	120-174	-16.59	10.66	22 10 01	
	19 11 25	- 27 02	120-174	-16.63	10.68	22 10 12	
	19 11 37	- 27 02	120-174	-16.67	10.70	22 10 24	
	19 11 49	- 27 02	120-175	-16.71	10.71	22 10 36	
	19 12 01	- 27 02	120-175	-16.75	10.73	22 10 48	
	19 12 13	- 27 02	120-176	-16.79	10.75	22 11 00	
	19 12 25	- 27 02	120-176	-16.83	10.77	22 11 12	
	19 12 49	- 27 02	120-176	-16.87	10.79	22 11 24	
	19 13 00	- 27 02	120-177	-16.92	10.80	22 11 36	
	19 13 12	- 27 02	120-177	-16.96	10.82	22 11 47	
	19 13 36	- 27 02	120-178	-17.04	10.85	22 11 59	
	19 13 48	- 27 02	120-178	-17.08	10.87	22 12 11	
	19 14 00	- 27 02	120-179	-17.16	10.89	22 12 23	
	19 14 12	- 27 02	120-179	-17.20	10.91	22 12 35	
	19 14 24	- 27 02	120-179	-17.24	10.93	22 12 47	
	19 14 36	- 27 03	120-180	-17.28	10.96	22 12 59	
	19 15 00	- 27 03	120-180	-17.37	11.00	22 13 23	
	19 15 12	- 27 03	120-181	-17.41	11.03	22 13 47	
	19 15 24	- 27 03	120-181	-17.45	11.05	22 14 11	
	19 15 36	- 27 03	120-181	-17.49	11.05	22 14 35	
	19 16 00	- 27 03	120-182	-17.53	11.07	22 14 47	
	19 16 12	- 27 03	120-183	-17.57	11.09	22 14 59	
	19 16 36	- 27 03	120-183	-17.65	11.14	22 15 11	
	19 16 48	- 27 03	120-184	-17.74	11.16	22 15 35	
	19 17 00	- 27 03	120-184	-17.78	11.17	22 15 47	
	19 17 12	- 27 03	120-184	-17.82	11.19	22 15 59	
	19 17 24	- 27 03	120-185	-17.86	11.21	22 16 10	
	19 17 36	- 27 03	120-185	-17.90	11.23	22 16 22	
	19 17 48	- 27 03	120-185	-17.94	11.24	22 16 34	
	19 18 00	- 27 03	120-186	-17.98	11.26	22 16 46	
	19 18 12	- 27 03	120-186	-18.03	11.28	22 16 58	
	19 18 35	- 27 03	120-187	-18.06	11.30	22 17 09	
	19 18 47	- 27 03	120-187	-18.15	11.33	22 17 33	
	19 18 59	- 27 03	120-188	-18.19	11.35	22 17 45	
	19 19 11	- 27 03	120-188	-18.27	11.37	22 17 57	
	19 19 35	- 27 03	120-188	-18.31	11.38	22 18 09	
	19 19 47	- 27 03	120-190	-18.35	11.40	22 18 21	
	19 19 59	- 27 03	120-190	-18.39	11.42	22 18 33	
	19 20 23	- 27 03	120-190	-18.44	11.44	22 18 45	
	19 20 35	- 27 03	120-190	-18.48	11.47	22 19 09	
	19 20 47	- 27 03	120-191	-18.56	11.49	22 19 21	
	19 21 11	- 27 03	120-191	-18.60	11.52	22 19 45	
	19 21 23	- 27 03	120-192	-18.64	11.54	22 20 09	
	19 21 35	- 27 03	120-192	-18.72	11.58	22 20 21	

Handwritten annotations: "Aug 15, 1977" (upper right); "Wow!" (left margin, with arrow). Circled column values in the intensity grid: 6 E Q U J 5.

terior View of Cell House, new Illinois State Penitentiary at Stateville, near Joliet,

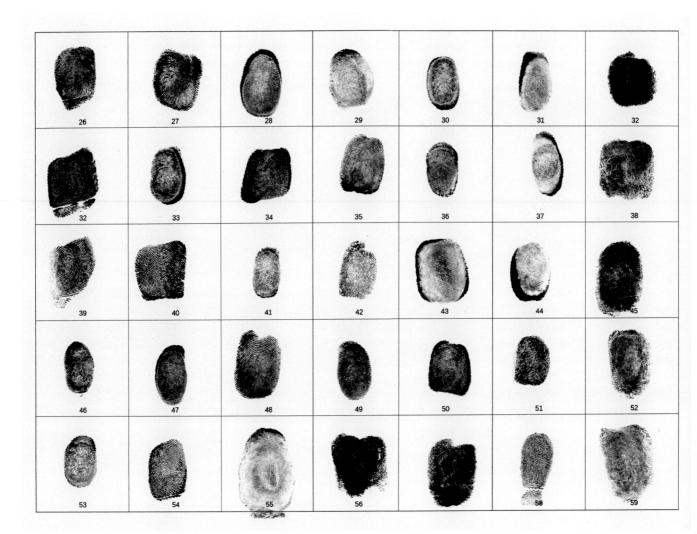

সেবাইতমোখ্যমুত একরাউ মহাসেবাসাহেব—
বরাবরেষু—

বিনয় পূর্ব্বা রক্ষাবর লেখাই সবিনিষ্কুত সবব
সুমাপুর কিন্তু সাহাট গণ মিনতে কসৃ নক্ষাবতে—
আমি আপনকার নিকটি এই সাদের লাখিযা দিতাহি—
যেমো রুখনাথ সবের কীবতি ২০০০ কৃ হাজার—
কার ঝেল মুটি অনতিযাদিবে এবং কি সবতকরা
৪, তাবানিনার হিসাবে, আপনকার নিকটি সবব
মুকিনে আপনকার নিকটি হুবতে কিং ১৫, গোনার
ষকা লাইয়া সাবে লাখিযা দিলাম, ই তি সবকাবা
একবাইখাব মুকাঈ মো সনংতততসবীতনসকনত্তাবেনী

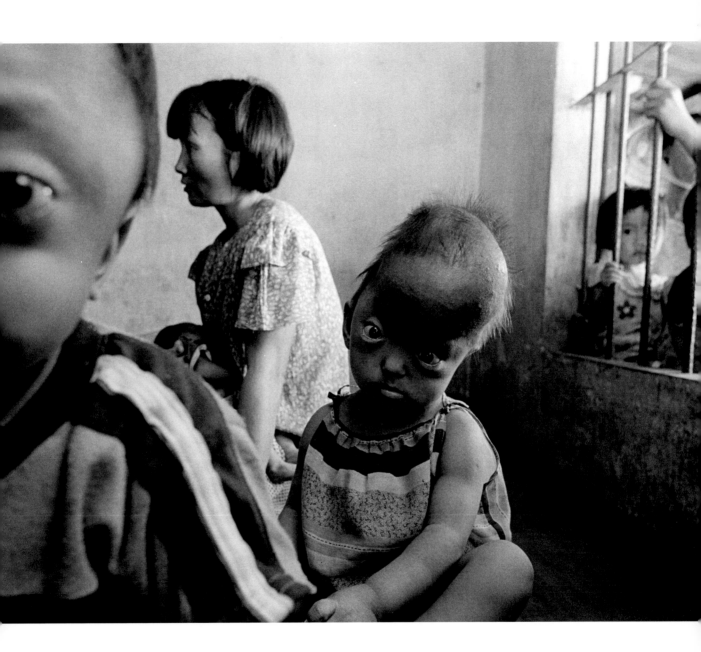

Oath of Allegiance.

I, _Ludwig Josef Johann Wittgenstein_
swear by Almighty God that I will be faithful and bear true allegiance to His Majesty,
King George the Sixth, His Heirs and Successors, according to law.

(Signature) _(Ludwig) Josef Johann Wittgenstein_

Sworn and subscribed this _12_ day of _April_ 193_9_, before me,

(Signature) _A. E. Ayres_

Justice of the Peace for ....

A Commissioner for Oaths.

Address { _1 Guildhall Chambers_
Basinghall Street London EC 2

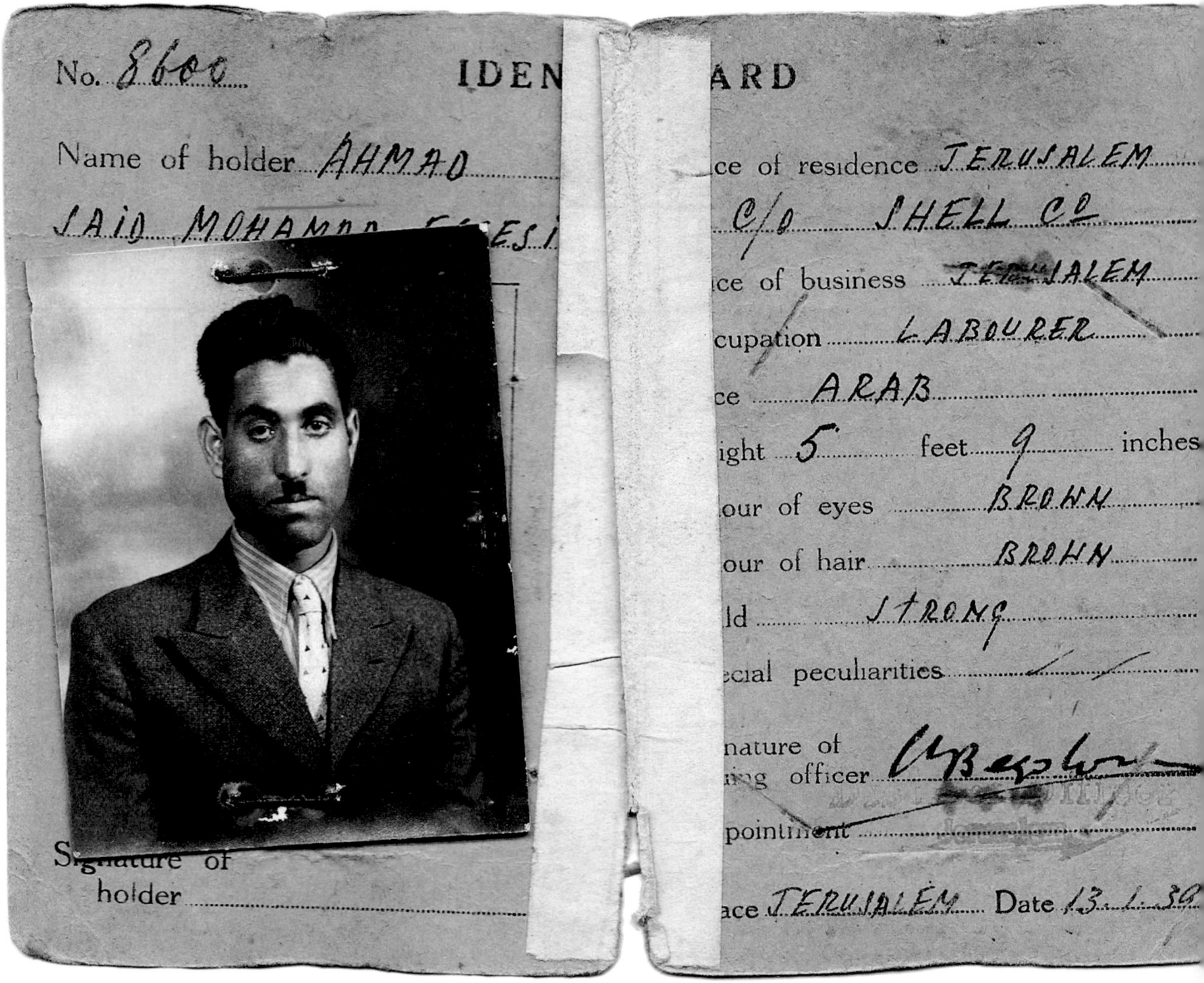

No. 8600   IDEN[TITY C]ARD

Name of holder AHMAD
SAID MOHAMAD FUESI[...]

[Pla]ce of residence JERUSALEM
  C/O SHELL Co

[Pla]ce of business JERUSALEM

[Oc]cupation LABOURER

[Ra]ce ARAB

[He]ight 5 feet 9 inches

[Col]our of eyes BROWN

[Col]our of hair BROWN

[Bui]ld STRONG

[Sp]ecial peculiarities —

[Sig]nature of
[issu]ing officer ABerglo[...]

[Ap]pointment —

[Pl]ace JERUSALEM   Date 13. 1. 39

Signature of
holder

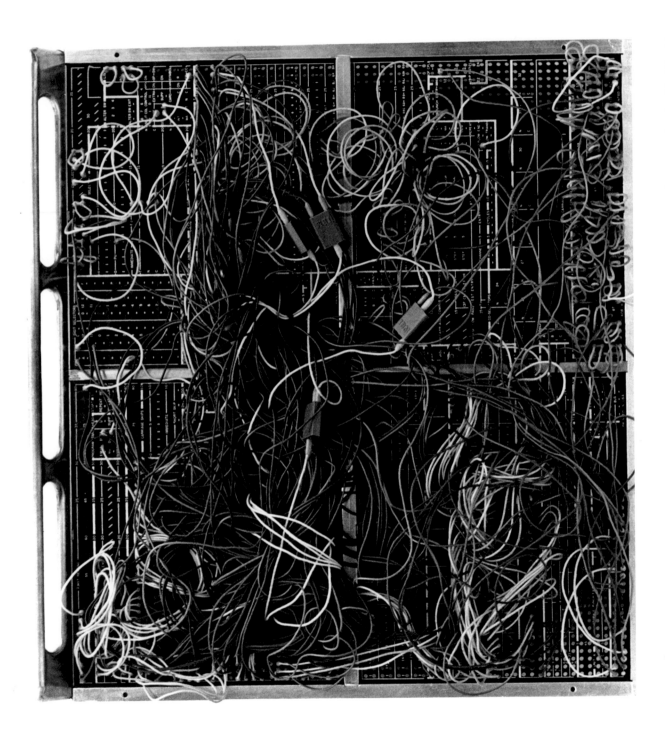

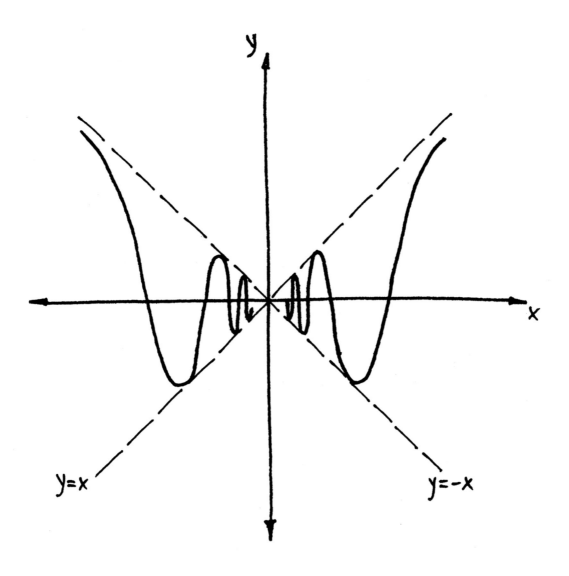

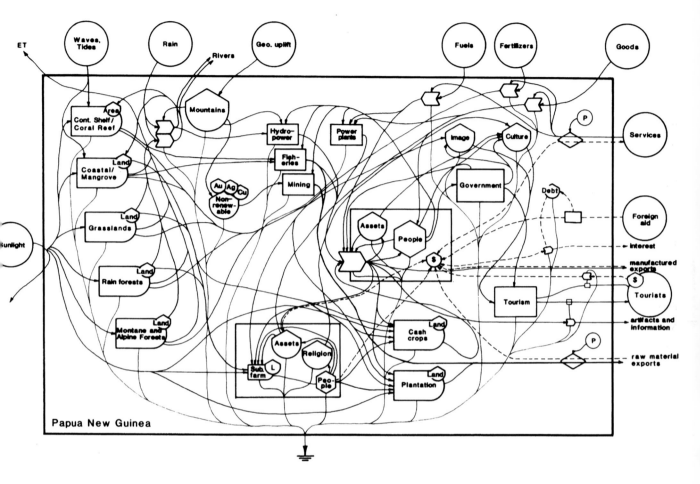

ET

Waves, Tides

Rain

Geo. uplift

Fuels

Fertilizers

Goods

Rivers

Mountains

Area Cont. Shelf / Coral Reef

Hydro-power

Power plants

Image

Culture

P

Services

Coastal/ Mangrove

Land

Fish-eries

Mining

Au Ag Cu Non-renew-able

Government

Debt

Foreign aid

Grasslands

Land

Assets

People

$

interest

Sunlight

manufactured exports

$

Tourists

Rain forests

Land

Tourism

Land Montane and Alpine Forests

Cash crops

Land

artifacts and information

P

Assets

Religion

Sub farm

L

People

Plantation

Land

raw material exports

Papua New Guinea

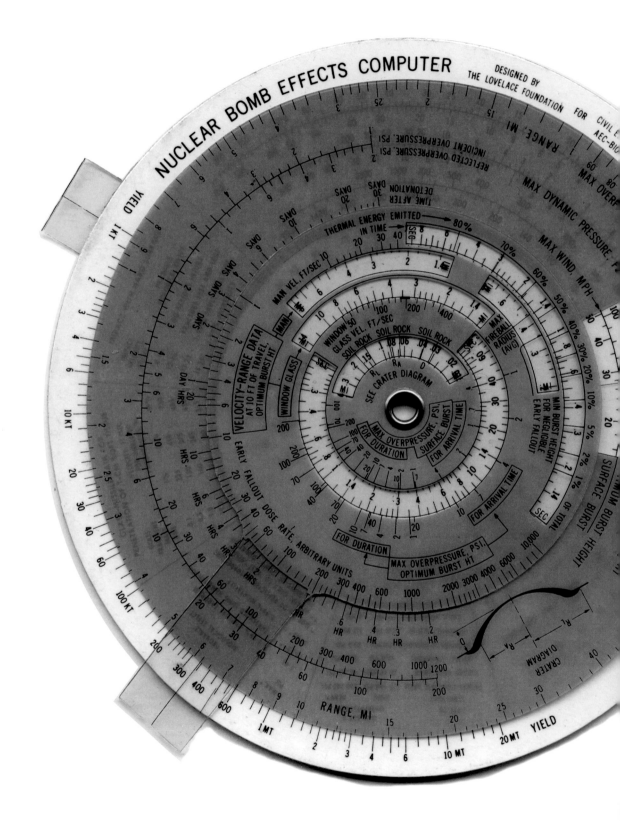

CONTRACT
(29-1)1242

BASED ON DATA FROM
"THE EFFECTS OF NUCLEAR WEAPONS".

REVISED EDITION
1962

SEA-LEVEL CONDITIONS)

(COMPUTED FOR

OPTIMUM BUR

3

2.5

2

1.8

1.6

1.4

1.2

1

Ontological proof
(*1970)

Feb. 10, 1970

$P(\varphi)$ φ is positive   (or $\varphi \in P$).

Axiom 1. $P(\varphi).P(\psi) \supset P(\varphi.\psi)$.[1]

Axiom 2. $P(\varphi) \vee P(\sim\varphi)$.[2]

Definition 1. $G(x) \equiv (\varphi)[P(\varphi) \supset \varphi(x)]$ (God)

Definition 2. $\varphi \, \mathrm{Ess.} \, x \equiv (\psi)[\psi(x) \supset N(y)[\varphi(y) \supset \psi(y)]]$.   (Essence of x)[3]

$$p \supset_N q \;=\; N(p \supset q). \quad \text{Necessity}$$

Axiom 3. $\begin{aligned} &P(\varphi) \supset NP(\varphi) \\ &\sim P(\varphi) \supset N\sim P(\varphi) \end{aligned}$

because it follows from the nature of the property.[a]

Theorem. $G(x) \supset G \, \mathrm{Ess}.x$.

Definition. $E(x) \equiv (\varphi)[\varphi \, \mathrm{Ess} \, x \supset N(\exists x)\,\varphi(x)]$.   (necessary Existence)

Axiom 4. $P(E)$.

Theorem. $G(x) \supset N(\exists y)G(y)$;
hence $(\exists x)G(x) \supset N(\exists y)G(y)$;
hence $M(\exists x)G(x) \supset MN(\exists y)G(y)$.   ($M$ = possibility)
$M(\exists x)G(x) \supset N(\exists y)G(y)$.

$M(\exists x)G(x)$ means the system of all positive properties is compatible.   2
This is true because of:
Axiom 5. $P(\varphi).\varphi \supset_N \psi :\supset P(\psi)$, which implies

$$\begin{cases} x = x & \text{is positive} \\ x \neq x & \text{is negative.} \end{cases}$$

[1] And for any number of summands.

[2] Exclusive or.

[3] Any two essences of x are *necessarily equivalent*.

[a] Gödel numbered two different axioms with the numeral "2". This double numbering was maintained in the printed version found in *Sobel 1987*. We have renumbered here in order to simplify reference to the axioms.

But if a system S of positive properties were incompatible, it would mean that the sum property s (which is positive) would be $x \neq x$.

Positive means positive in the moral aesthetic sense (independently of the accidental structure of the world). Only then [are] the axioms true. It may also mean pure "attribution"[4] as opposed to "privation" (or *containing* privation). This interpretation [supports a] simpler proof.

If φ [is] positive then *not*: $(x)N\sim\varphi(x)$. Otherwise: $\varphi(x) \supset_N x \neq x$; hence $x \neq x$ [is] positive, so $x = x$ [is] negative, contrary [to] Axiom 5 or the existence of positive properties.

[4] I.e., the disjunctive normal form in terms of elementary properties[b] contains a member without negation.

[b] Here Gödel uses the abbreviation "prop.", which could be read, in isolation, either as "properties" or "propositions". In the context, however, it is clear that it is properties whose positiveness is under discussion. The related discussion in the excerpts from "Phil XIV" in the appendix, below, explicitly concerns "positive properties". With regard to fn. 4, where the reference to "disjunctive normal form" might lead us to think first of propositions, note that in "Phil XIV", p. 108, Gödel speaks explicitly of properties ("Eigenschaften") that are "members of the conjunctive normal form" of complex properties. An interpretation of fn. 4 is offered in the introductory note, pp. 397-398 above.

17173	92771	28940	99691	30884	12991	04206	892
68552	66848	00342	47015	41001	23628	26246	964
87924	31942	48589	69215	43143	56072	54112	518
66717	92224	50359	85655	08702	81146	13574	882
			04581	35205	31450	88812	944
60129	01481	01705	97668	17576	13709	47285	168
47338	17139	77798	34271	74510	93251	63084	632
50204	12103	29859	33980	49492	03231	92102	541
29722	50186	77468	36427	50460	48901	51758	508
94014	01210	80086	25041	90722	13080	66898	2128
39227	51911	52129	02563	07780	03797	09722	152
09432	70694	70627	46930	74591	70209	60344	769
94817	39324	94127	70930	22585	25175	01669	646
02216	39545	65795	24918	09444	27483	26939	6776
5510	75880	32003	15648	08831	40362	96956	265
01423	22408	88501	97336	10337	99704	83849	013
48233	75410	58823	02608	86138	92642	11707	1140
95287	48500	35499	44472	43915	09708	26441	9766
5988	98120	25630	85345	43924	24414	39029	3533
85996	11692	45985	72051	39989	49191	73607	8300

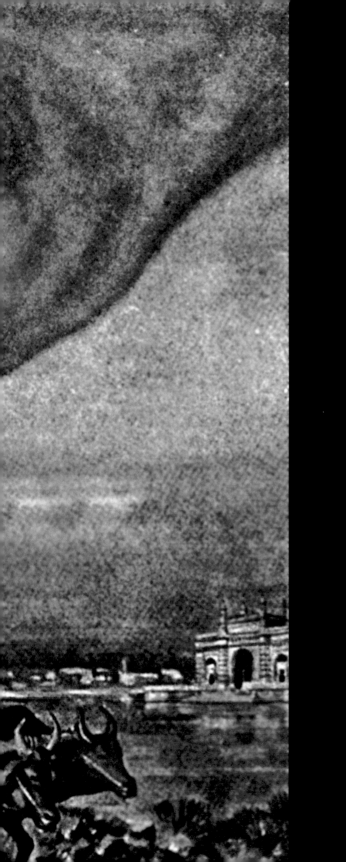

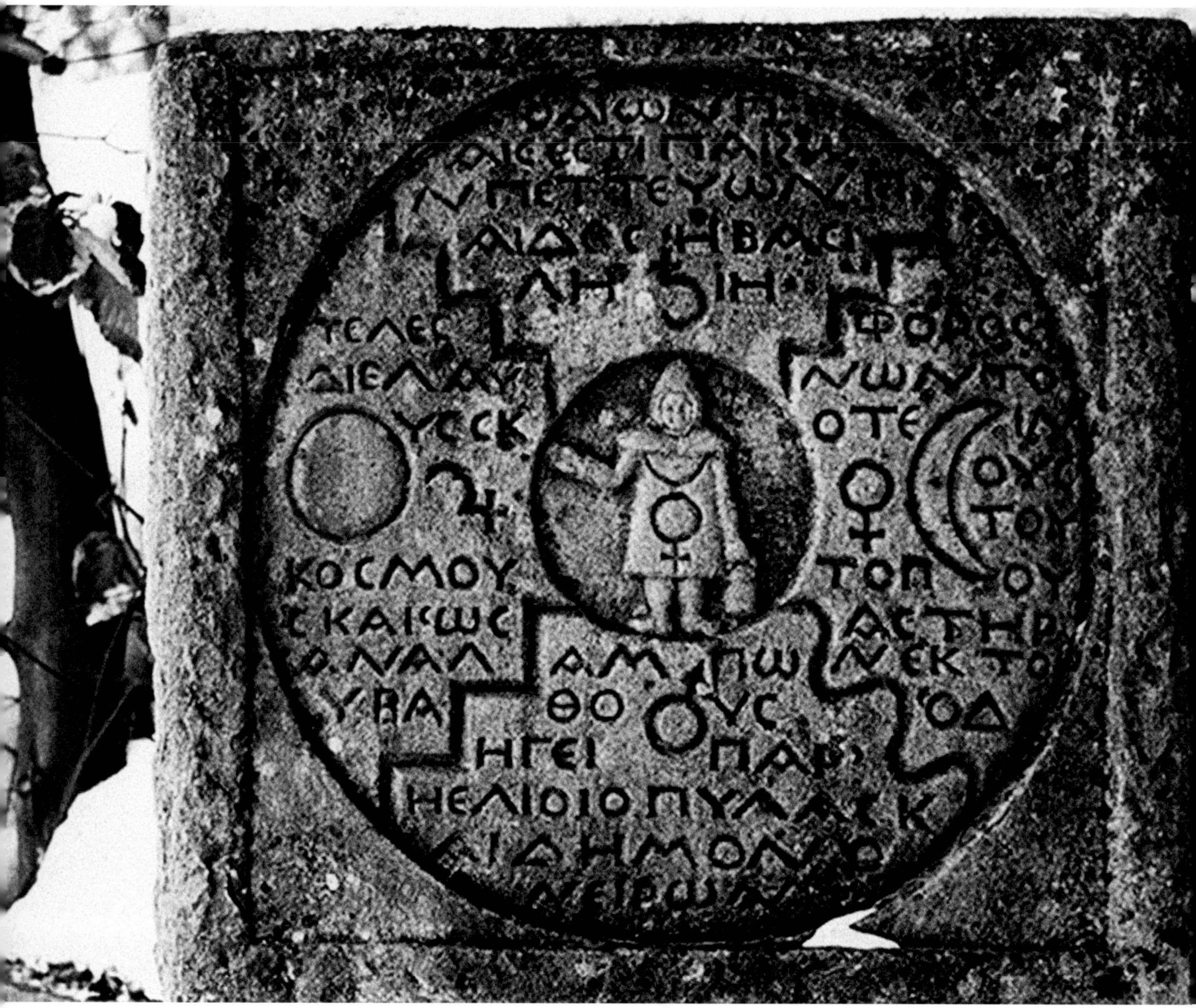
ΤΕΛΕϹ
ΚΟϹΜΟΥ
ΟΤΕ
ΤΟΠ
ΘΟ
ΗΓΕΙ
ΠΑΡ
ΗΕΛΙΟΙΟ ΠΥΛΑ

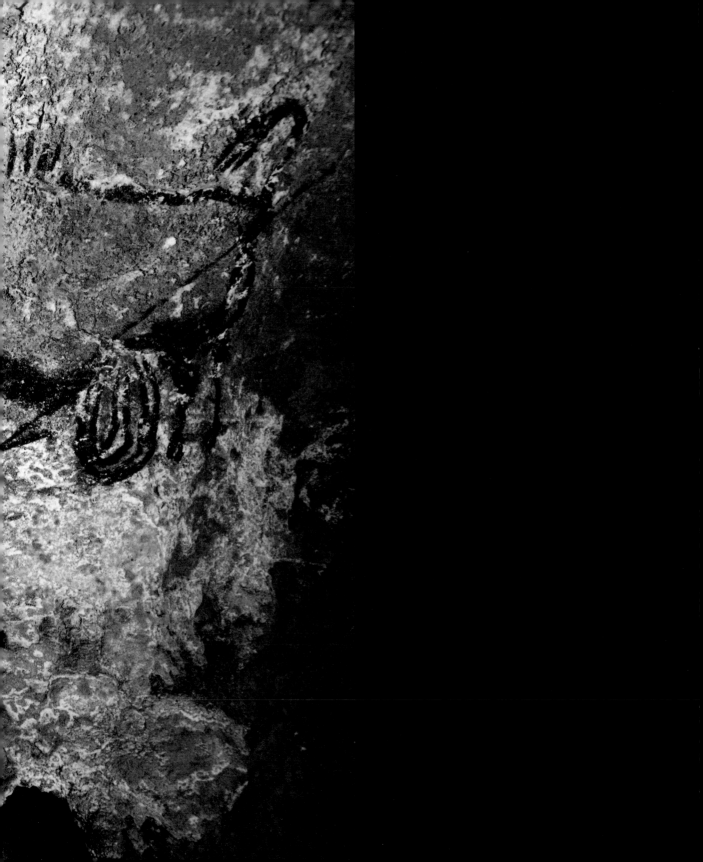

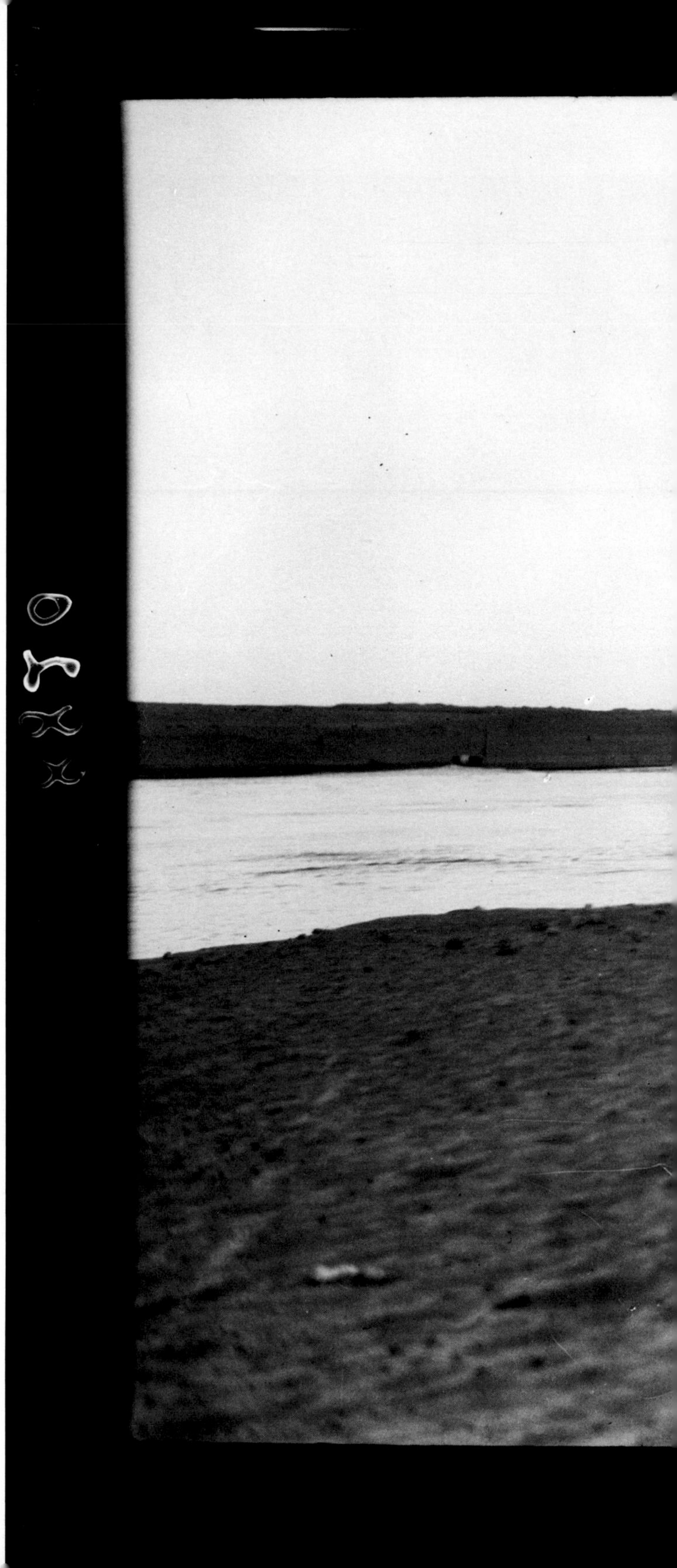

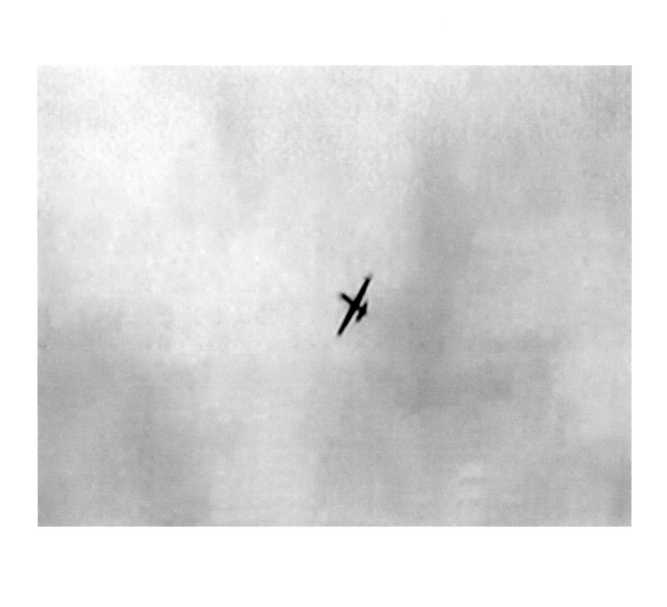

NOTES ON THE ONE HUNDRED PICTURES

1. BACK OF PAUL KLEE'S *ANGELUS NOVUS* DRAWING

> My wing is poised to beat
> but I would gladly return home
> were I to stay to the end of days
> I would still be this forlorn[1]

There is a painting by Klee called *Angelus Novus*. It shows an angel who seems about to move away from something he stares at. His eyes are wide, his mouth is open, his wings are spread. This is how the angel of history must look. His face is turned toward the past. Where a chain of events appears before *us, he* sees one single catastrophe, which keeps piling wreckage upon wreckage and hurls it at his feet. The angel would like to stay, awaken the dead, and make whole what has been smashed. But a storm is blowing from Paradise and has got caught in his wings; it is so strong that the angel can no longer close them. This storm drives him irresistibly into the future to which his back is turned, while the pile of debris before him grows toward the sky. What we call progress is *this* storm.[2]

1. Gershom Scholem, "Greetings from Angelus," in *The Fullness of Time*, trans. Richard Sieburth (Jerusalem: Ibis Editions, 2003).
2. Walter Benjamin, "On the Concept of History," in *Selected Writings*, Vol. 4, trans. Harry Zohn, ed. Michael W. Jennings (Cambridge: Harvard University Press, 2003), 392–93.

2. SOYUZ FG ROCKET LAUNCH, BAIKONUR COSMODROME, KAZAKHSTAN

3. TYPHOON, JAPAN, EARLY TWENTIETH CENTURY

4. GREEK AND ARMENIAN ORPHAN REFUGEES EXPERIENCE THE SEA FOR THE FIRST TIME, MARATHON, GREECE

5. EARTHRISE

6. OLD OPERATING THEATER, ST. THOMAS CHURCH, SOUTHWARK, LONDON

The Old Operating Theater in the garret of St. Thomas Church, London, was built in 1822 to provide surgical facilities for the women's ward at St. Thomas Hospital. The high vaulted ceiling and skylight in the garret's roof ensured that the theater received ample light, making it easy for students to observe proceedings from the surrounding bleachers.

Operating theaters of this sort, designed to optimize viewing, gained prominence in the nineteenth century.

7. GLIMPSES OF AMERICA, AMERICAN NATIONAL EXHIBITION, MOSCOW WORLD'S FAIR

Charles and Ray Eames's 1959 film, *Glimpses of the U.S.A.,* was commissioned by the U.S. State Department for the National Exhibition in Moscow. This seven-screen installation flashed 2,200 images (both still and moving) inside a massive golden geodesic dome that served as a major throughway during the event. That summer, over two million people visited the display, which was the first cross-cultural exchange between the two nations since the Bolshevik Revolution.

Glimpses of the U.S.A. popularized the idea of watching an array of television screens at the same time—a way of seeing that quickly spread to NASA mission control centers, air traffic control towers, factories, financial centers, and the multiple monitors found on many computer desktops.

8. CHEYENNE MOUNTAIN, COLORADO SPRINGS, COLORADO

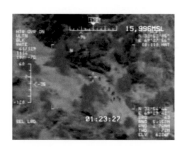

9. MIGRANTS SEEN BY PREDATOR DRONE, U.S.-MEXICO BORDER

10. ZORK

11. ENTANGLED BANK

12. ELECTRON MICROSCOPIC PHOTOGRAPH, MARTIAN METEORITE

On August 7, 1996, President Bill Clinton convened a press conference to specifically address this photograph, which shows the view through an electron microscope of a Martian meteorite found in Antarctica. The image seems to provide evidence of fossilized microbes embedded in the Martian rock—evidence for the existence of life on Mars. Clinton claimed the image "might constitute one of the most stunning insights into our universe that science has ever uncovered," one whose implications "are as far-reaching and awe-inspiring as can be imagined."

For anthropologist Stefan Helmreich, however, the photograph is a reflection of our own fantasies: "The image itself needs so much interpretation—is it a giant worm? Is it a close-up of cheese from the 1950s? Is it, as some suggest, a fossil nanobe . . . ? It brings *something* into relief, but what that is remains unclear."

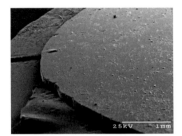

13. HELA CELLS, CARNEGIE MELLON UNIVERSITY, PITTSBURGH, PENNSYLVANIA

On January 29, 1951, Henrietta Lacks walked out of Baltimore's Johns Hopkins Hospital with a test tube filled with radium sewn inside of her—aggressive treatment, she'd been told, for an aggressive form of cervical cancer. She hadn't been told that during the procedure, two dime-sized slices of tissue were cut from the lining of her cervix. Lacks, a working class, African-American tobacco farmer, died eight months later at the age of thirty-one. But her tissue samples lived on in a hospital laboratory. Lacks's cells were the first "immortal" human cells ever discovered, and they are still alive today.

Since then, thousands of metric tons of HeLa cells (as they were first codenamed to protect Lacks's identity) have been produced for medical research. NASA sent them into orbit to see what would happen to human cells in zero gravity. They were also key in discovering a polio vaccine and treatments for Parkinson's disease. "Though those cells have done wonders for science, Henrietta—whose legacy involves the birth of bioethics and the grim history of experimentation on African-Americans—is all but forgotten," writes journalist Rebecca Skloot.

Lacks left behind five children; none knew that their mother's cells had been preserved until scientists approached them twenty-five years later, wanting their cells for research, too. Although Henrietta Lacks's cells enabled countless medical advances, generating billions of dollars for pharmaceutical companies, Lacks's family continued living in poverty. In 2010, Lacks's unmarked grave received a headstone with an epitaph written by her grandchildren, "In loving memory of a phenomenal woman, wife and mother who touched the lives of many. Here lies Henrietta Lacks (HeLa). Her immortal cells will continue to help mankind forever. Eternal Love and Admiration, From Your Family"

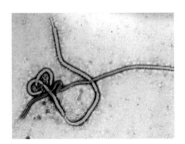

14. TRANSMISSION ELECTRON MICROGRAPH, EBOLA VIRUS

15. LEON TROTSKY'S BRAIN, MEXICO CITY

16. DR. EDWARD J. TRIPLETT TAPING PICTURE (TELEVISION)

This image is of biology professor Edward L. Triplett, who taught at the University of California at Santa Barbara for thirty-eight years before passing away in 2007. It is not clear what Triplett is demonstrating or drawing in this photograph.

The photograph has an exceptional history. "In 1963, the president of the University of California system, Clark Kerr, commissioned renowned artist Ansel Adams to take photographs of the university's future," explains UC Berkeley's Professor Catherine Cole, "to project, as much as possible, the next hundred years." It was unusual commission: take a picture of the future. Kerr's underlying assumption was that the future is made, in part, by allowing ourselves to imagine what it could be, and going ahead and implementing that vision. Ansel Adams was hired to produce 1,000 images. He turned in 6,700. The *Fiat Lux* collection is now owned by the UC regents.

In 2010, when the University of California was at a "major crossroads moment" in history, UC president Mark Yudof presided over a Commission on the Future. Its major conclusion was that "The future cannot be avoided."

17. WHALE SHARK, GEORGIA AQUARIUM

On his eighteenth birthday, Zac Wolf photographed a whale shark at the Georgia Aquarium in Atlanta—at the time the largest aquarium in the world, and the first American tank to keep the huge animal in captivity. "I don't shoot fish," says Wolf, a wedding photographer. "I shoot people. But I liked the way it moved through the water, and the way the people moved as they watched the shark." This photograph was seminal to Wolf's career: "[It] got me to say, I can really do this."

The Georgia Aquarium acquired two whale sharks in 2005—both shipped from Taiwan, where they had been accidentally caught in waters off Japan's eastern coast. The sharks died within six months of each other, most likely of a gastrointestinal disease.

Four new whale sharks live now live in the tank. "You can probably find a similar image easily enough," Wolf says. "But you can never shoot that same one again. This one shows scale really well."

18. SHOP WINDOW AND TAILOR'S DUMMIES

19. FILMING *CONQUEST OF THE PLANET OF THE APES* (PRODUCTION STILL), UNIVERSITY OF CALIFORNIA

20. GAS MASKS, WORLD WAR I

21. STEALTHY INSECT SENSOR PROJECT, LOS ALAMOS, NEW MEXICO

Scientists at Los Alamos' Stealthy Insect Sensor Project have been training honeybees since early 2006 to detect explosives for the Defense Advanced Research Projects Agency (DARPA).

To train the bees, researchers strap them into thin tubes and bombard them with six-second blasts of vapor containing TNT, TATP, C4, and other explosives. A sugar-water reward is mixed into the last three seconds of the blast. After just four rounds of this training, the honeybees learn to associate trace explosives with the sugary reward, and they begin sticking out their proboscis (tongue) in anticipation when they detect the scent of explosives. Once the bees have been trained, they are placed in a specially designed portable bomb-detecting device and become living bomb sensors.

A Los Alamos promotional video for the Stealthy Insect Sensor Project describes the honeybees as "nature's rugged robots."

Anthropologist Joseph Masco has a more reflective take: "In the twenty-first century, humans will be forced to recognize that there is no nature anymore, only the cumulative effects of human activities on the biosphere and the mutable forms that are consciously, and unconsciously, being produced both inside and outside the laboratory. These militarized bees are thus a true emblem of our age, an era when life itself is redesigned in the name of protecting the sanctity of life."

22. MOTH FOUND IN MARK II AIKEN RELAY CALCULATOR, HARVARD UNIVERSITY

The Mark I computer, designed in 1944 by Harvard University and IBM, was the first programmable digital computer made in the United States. A solution to the growing need for reliable data for artillery firing tables, the Mark I was also capable of writing equations for the diminishing number of competent mathematicians. It was eight feet tall and fifty-one feet long, weighed five tons, and contained five hundred miles of wires, switches, relays, shafts, and clutches.

This image provides an apocryphal explanation for the origin of the computer bug: U.S. Navy Captain Grace Murray Hopper, an engineer who helped shepherd the development of the computer during WWII, remembered working late on the evening of September 9, 1947, while testing a next-generation Mark II computer. "Things were going badly," she recounted in a lecture at Long Island University, ten years before she died in 1992. "There was something wrong in one of the circuits of the long glass-enclosed computer. Finally, someone located the trouble spot and, using ordinary tweezers, removed the problem, a two-inch moth." It had been blocking the reading of holes in the computer's paper tape. The engineers reported that the Mark II had been "debugged."

"From then on, whenever anything went wrong with a computer we said it had bugs in it," Hopper said.

23. WARNING FROM SPACE, DAIEI FILMS

24. HALL OF BULLS, LASCAUX CAVE

25. DICTIONARY OF VOLAPUK

In 1879 in Baden, Germany, Father Johann Martin Schleyer created a universal language at the behest of God, speaking to Schleyer in a dream. He called this new language Volapük or World Speak. Volapük was a simple language meant to give Catholic readers from different linguistic backgrounds an easier time reading aloud from the Bible. Most of the twenty thousand words in the Volapük vocabulary are monosyllabic. Schleyer also omitted the letter "r" from this language as well as the sounds "th" and "ch," which are difficult pronunciations for most non–English speakers.

Within ten years nearly one million people were conversing in the language. Volapük-specific publications were widely available, textbooks about the language were published in twenty–five languages, and Volapük societies proliferated across Europe.

Yet Volapük's popularity as a universal language was eclipsed by the rise of Esperanto in the early twentieth century. Esperanto expressions began mocking Volapük; in Esperanto "that sounds like Greek to me" became "that sounds like Volapük to me," and "Volapukajxo" is a synonym for nonsense.

In 2000, there were twenty to thirty Volapük speakers on Earth.

26. *TOWER OF BABEL* (DETAIL), PIETER BRUEGEL THE ELDER

27. DEMONSTRATION OF EATING, LICKING, AND DRINKING

When George Helou (pictured on the right) arrived to study astronomy at Cornell University in 1975, the twenty-three-year-old Lebanese graduate student found what he described as an "American experience." Although Helou had fallen in love with the night sky in his home country, he never imagined studying astronomy could become a career. When Carl Sagan and other Cornell faculty began developing the Voyager space probes' Golden Records, Helou volunteered to help. The idea that he could choose whether or not to work on his professors' projects was as alien to him as the idea that there would not be an exam. "The whole thing, including the notion that we should send pictures to extraterrestrials, all seemed very 'American,'" he recalled Nonetheless, he enjoyed the "exercise in self-assessment more than anything else . . . How do you describe humanity? And without any cultural references?"

Helou joined Carl Sagan's assistant Wendy Gradison and Argentine radio-astronomer Val Boriakoff to create an image intended to be a "demonstration of eating, licking, and drinking." The first two activities proved easy enough: Gradison was given an ice cream cone and Boriakoff a toasted tuna sandwich (although he hated tuna). But showing the act of drinking proved to be a challenge. Their first attempts "didn't show much," said Helou. "I suggested we use a jug with a special spout. Because if you really want to capture drinking, you have to show water flowing into the mouth. And from a scientific perspective, that was our first concern—to show the flow." The budding scientist also wanted to indicate that water has a surface and is subject to gravity, and that gravity has a particular strength. "We took numerous takes over a twenty-minute period before we got it right. And I think we did."

Helou went on to join the faculty at the California Institute of Technology, where he has spent the past twenty-nine years. Boriakoff discovered PSR 1953+29, the first binary millisecond pulsar, in 1983. He passed away in 1999. We were unable to locate Wendy Gradison for this book.

28. THE GREAT WALL OF CHINA

29. *AN ESSAY TOWARDS A REAL CHARACTER AND A PHILOSOPHICAL LANGUAGE*, JOHN WILKINS

John Wilkins was a seventeenth-century British polymath who wrote on astronomy, cryptography, engineering, theology, and the orbit of an invisible planet. He was the first secretary of the Royal Society of London.

Wilkins' most well-known book, the 1668 *An Essay Towards a Real Character and a Philosophical Language*, is an attempt to develop a universal language. An unambiguous replacement for Latin, he reasoned, would clear up a number of philosophical problems, many of which could be reduced to linguistic quirks. Furthermore, a universal language would be useful to merchants, diplomats, and scholars.

Although *An Essay Towards a Real Character* preceded Linnaeus's *Systema Naturae* by nearly a century, Wilkins' proposed vocabulary is based on taxonomies. He divides the universe into sixteen genera: Transcendental, Discourse, God, World, Element, Stone, Metal, Herbs, Shrub, Tree, Animals, Parts, Quantity, Quality, Action, and Relation. Each genus is signified by two-letter root word; specific concepts or words are formed by adding letters to each root. For example, the root word De signifies the genus "element"; Deb is "fire"; Det is "appearing meteor"; Deta is "rainbow"; Deta is "halo."

In an essay about Wilkins' work, Jorge Luis Borges wrote that the "ambiguities, redundancies and deficiencies . . . remind us of those which doctor Franz Kuhn attributes to a certain Chinese encyclopedia entitled 'Celestial Empire of benevolent Knowledge.' In its remote pages it is written that the animals are divided thusly: (a) belonging to the emperor, (b) embalmed, (c) tame, (d) sucking pigs, (e) sirens, (f) fabulous, (g) stray dogs, (h) included in the present classification, (i) frenzied, (j) innumerable, (k) drawn with a very fine camelhair brush, (l) et cetera, (m) having just broken the water pitcher, (n) that from a long way off look like flies."

"It is clear that there is no classification of the universe not being arbitrary and full of conjectures," concluded Borges. "The reason for this is very simple: we do not know what thing the universe is."

30. AI WEIWEI, STUDY OF PERSPECTIVE—EIFFEL TOWER, 1995–2010

31. NAGOYA PUBLIC AQUARIUM

One weekend in 2007, Paul Synnott and his wife, Koyuki, took a shinkansen (bullet train) from Izumi City to Nagoya to hear her favorite singer perform in a shopping mall. Afterward, they went to the Nagoya Public Aquarium to see the orcas, also known as killer whales. "This woman, a member of the staff, bowed to the whale before and after feeding her," Synnott recalls. "It was a common ritual, and one of many incredible moments during my stay in Japan."

Synnott later learned that the orca was Ku, who died on September 19, 2008, at age sixteen. He remembers how the feeder felt Ku's skin for tumors that covered her body. Ku had been caught in 1997 off the Hatajiri Bay in Taiji—a southern Japanese town made famous in *The Cove*, an American documentary that questioned the ethics of local dolphin hunts—and sold to the Taiji Whale Museum for $250,000. After her health deteriorated, the museum eventually loaned her to the Nagoya Public Aquarium. Ku died of heart failure in a medical pool; she had been diagnosed with herpes shortly before her death.

Synnott had moved to Japan after college to teach English and to be near Koyuki, his pen pal. "We met online, and then started writing real letters," he says. Paul and Koyuki now live in his hometown, County Antrim, in Northern Ireland.

32. OCCUPY HONG KONG

When the global Occupy movement first arose in the fall of 2011, the Hong Kong demonstrations, held on the grounds of the city's International Financial Center, were largely peaceful. Brett Elmer, a freelance photographer hired to shoot the crowds for Time Out Hong Kong, watched as police and picketers made small talk, standing beneath the black and yellow flags of the People's Power Party, the pro-labor group that organized the event. "Locals tend to be more intellectual in their critique, since they've been free to voice their opinions, unlike citizens of mainland China," Elmer said. "I was quite surprised by the turnout. I never expected such a crowd."

Over one thousand people attended the rally—what Elmer described as the "height of the global protest movement in Asia"—holding signs with slogans such as "Banks are cancers" and "Stop the poisoning." Unemployment rates have been relatively low in Hong Kong, where the economy is primarily fueled by financial sector jobs. Still, the driving issue has been the exponential rise in the cost of living and the widening gap between income levels. "It's a well-known fact that a small group of tycoons control everything we do here," said Elmer, a transplant from Sydney, Australia.

33. CRISTINA LLANOS, BOLONIA BEACH

Cristina Llanos works as an artist under two names, her own and "Nikita Rodriguez." As the latter, she makes drawings and paintings (among other things) of skeletal figures with long slits for eyes and rows of small, pointy teeth. Her boyfriend, Aitor Mendez, is a graphic designer and activist. They both live in Madrid.

In 2010, the couple vacationed with Mendez's family on Bolonia Beach, a small oceanfront town near Cadiz, Spain, that famously hosts concrete ruins dating back to the Roman Empire. Bolonia once served as a major shipping port known for exporting garum, an extract made from fermented fish and wine. Yet the beach itself has remained unspoiled by tourism or commercial development. "It's virgin land," Mendez says.

Halfway through their trip, the family traveled north to El Cañuelo Beach. They sunbathed and goofed around until well past dark. On that particular day, the water was calm, but because the sand sloped into it at a steep angle, the waves broke hard. "Everyone had fun, fighting with the foam," Mendez says. "I saw Cristina and ran into the water with my camera."

34. CARLOS BURLE, MAVERICKS

At the 2010 Mavericks surfing competition near Half Moon Bay, California, Patrick Sanford waited patiently for Carlos Burle, a Brazilian big-wave surfer, to catch a giant wave. Sanford remembers, "I was on a boat and could barely stand because we were rocking ten to fifteen feet vertically, just from the wakes. But once I saw him go into that beautiful barrel, I started unloading the shots, hoping to catch anything at all. I guess I was in the right place at the right time."

Sanford, who photographs big wave surfing all over the world, reserved a space on that boat well in advance. No competition is a sure thing; the ocean swell has to hit a specific size first, and once it does, surfers from around the world have just forty-eight hours to find their way to the wave.

Speaking from his experience as a former surfer, Sanford explains the danger involved in big-wave surfing. "It's man against nature in every possible way. You can't predict the shape of a wave or how it will peak. These guys are tremendous athletes. They have no fear." He remembers Sion Milosky, a thirty-five-year-old Hawaiian who drowned at Mavericks in 2011 when he was held underwater through two consecutive waves. Sanford recalls, "They found him floating near the break."

"I watched Carlos drop into a sizable one—it really looked like the water was going to overtake him," says Sanford. "But then it doubled up, and he got tucked in there. My photos don't do justice to the moment. Some of these waves are nearly fifty feet high. It's terrifying."

According to big-wave surfing icon Laird Hamilton, "Global warming can create bigger storms, and has, and is going to . . . and that's great for me."

35. CLATSKANIE, OREGON

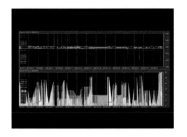

36. "CROP CIRCLES" IN FINANCIAL MARKET FORMED BY HIGH-FREQUENCY TRADING ALGORITHMS

Stock market quotes for each stock consist of four main components: bid price, bid size, ask price, and ask size. Any of these components can change in value. Around 2007, high-frequency trading (HFT) appeared as a new force for rapidly analyzing this data and seizing trading opportunities by using computer programs and specialized machines. Before HFT, it was rare to see quotes change more than a few times per second. Today, it is common to see five thousand or more quote changes per second in a single stock. This means that information is changing faster than it can travel. Someone outside the very small geographic area where these trades are happening will receive expired or incorrect information by the time the data reaches their computer.

When we plotted the four components in a line graph, revealing patterns emerged, which demonstrated HFT algorithms were underlying and controlling the market. The image here has two panels: the top panel is a plot of bid and ask prices, because they share a common scale, the bottom panel is a plot of bid and ask sizes, which also share a common scale. In the beginning, most patterns we found were simple: the price or size component would repeat over short cycles. Over time, more sophisticated patterns have appeared: the price or size component might have hundreds of unique values before repeating, for example.

By Eric Scott Hunsader, founder of Nanex, a company that tracks market events and phenomena using a real-time streaming data feed.

37. HYDRAULIC MINING, NEVADA COUNTY, CALIFORNIA

38. CCTV TOWER, BEIJING

39. BEAN FIELDS, SEABROOK FARM, BRIDGETON, NEW JERSEY

40. HEN FARM, SOUTH AUCKLAND

On a typical industrial poultry farm, chickens are housed in battery cages—tight, dark spaces that limit movement and maximize egg production. On Easter in 2007, members of the New Zealand–based animal rights group Open Rescue Collective each dressed as the Easter Bunny, broke into this hen farm, and released ten hens from their cages. They replaced them with vegan Easter eggs and a message:

"This year, the Easter Bunny explains that while eggs are traditionally a symbol of new life, this symbol has been perverted as hens kept in battery cages are the product of a lifetime of suffering and deprivation. We carried out this action to highlight the suffering of battery hens. Hens confined in battery cages are denied their most basic natural behaviors. They can't even walk or stretch their wings."

The rescued hens were relocated to single-family farms that house poultry and other animals raised for agriculture in clean open-air environments.

41. THE UNION STOCKYARD AND TRANSIT COMPANY, CHICAGO

42. CAT PIANO, ATHANASIUS KIRCHER

43. INDUSTRIAL FISHING OF ORANGE ROUGHY (DEEP SEA PERCH)

44. LERNAEAN HYDRA, ALBERTUS SEBA'S *THESAURUS*

The monstrous hydra is an engraved plate in Albertus Seba's *Thesaurus*, a treasury of natural curiosities published in 1734. A drawing of the hydra was sent to Seba, a Dutch apothecary and collector, by the burgomeister of Hamburg, then owner of the taxidermist–preserved creature. Seba's collection of shells, plants, insects, preserved birds, snakes, lizards, and exotic objects— specimens brought to him by merchants and sailors returning from afar in that age of "discovery"—was famous among both collectors and scientists.

In 1735, a young Carolus Linnaeus visited Seba twice; the collection of specimens would play a part in shaping Linnaeus's taxonomy of organic and inorganic life. Linnaeus, too, had monsters—creatures slightly set off from the norm. His 1735 *System Naturae* included genera of Troglodytes, Satyrs, Pygmies, and six monstrous varieties of *Homo sapiens*. But when Linnaeus visited Hamburg to examine in person the burgomeister's hydra, he declared the monster a fraud. It was composed of weasel parts and snakeskin, Linnaeus argued, probably made by monks to represent an apocalyptic beast. When the value of the hydra plummeted following Linnaeus's public exposure of the creature as a fraud, Linnaeus fled Hamburg, fearing retribution from the burgomeister, who had been offering the object for sale at an enormous sum.

By Katie Detwiler

45. DINOSAUR FOOTPRINTS, PALUXY RIVERBED, GLEN ROSE, TEXAS

46. CLONED TEXAS LONGHORNS

For many years, Starlight (on the right, with her clone, Starlight #4, and one of Starlight #4's calves) was the world's number one Texan Longhorn. In 2002, she became the world's first cloned or "scientifically reproduced" Longhorn, chosen for her astonishing 81⅝-inch tip-to-tip horn measurement.

Livestock cloning using somatic-cell nuclear transfer has been commercially practiced in the United States since 1998. In 2008, the Food and Drug Administration released a scientific risk assessment, concluding, "There is no indication that differences exist in terms of food safety for meat and milk of clones and their progeny compared with those from conventionally bred animals." The Longhorn cattle breeding community is required to label cows that have been produced using "scientific reproduction." Although most cloning technology is used in cattle breeding, cloned meat and the offspring of cloned animals, which are widely distributed in the food industry, don't have to be labeled.

Starlight currently lives with her family on the Clear Creek Pecan Plantation in Texas. Her caretaker, Joyce Fletcher, observes: "You will find this group functioning as a unit; where you see one, you will usually see all. I have come to understand this 'grouping' as a form of twin bonding." But, she cautions, "Applying the technology of cloning to any species as a tool of genetic improvement needs rigorous self-adjustment, for the seeds of human greed . . . are inevitable to grow in such a fertile field."

By Emily Parsons-Lord

47. GOLDSTONE DEEP SPACE COMMUNICATIONS COMPLEX, DEEP SPACE NETWORK, GOLDSTONE, CALIFORNIA

48. PERCIVAL LOWELL, MARTIAN CANALS

In 1894, astronomer Percival Lowell announced that he had discovered an extensive network of canals on Mars from his private observatory in Flagstaff, Arizona. Leading from the planetary poles to the dry expanses of arid land, Lowell observed sophisticated infrastructure designed to irrigate infertile land with the annual melt from the polar ice caps. According to Lowell, these canals were irrefutable evidence of an intelligent civilization on Mars and their last heroic effort to maintain life in an inhospitable climate. Though these canals remained unobserved by other astronomers, Lowell charted detailed maps of Mars and passionately held his conviction until his death in 1916.

Lowell had originally turned his telescope to Mars to investigate claims by the Italian astronomer Giovanni Schiaparelli that "canali" existed on Mars. The Italian word "canali," mistranslated into English as *canals* instead of *channels,* led English-speaking astronomers to expect nonnatural features on the surface of Mars, rather than naturally occurring continental channels.

Although *Mariner 4* flew by Mars in 1965 and conclusively disproved the existence of the canals, the question of what Lowell saw remained unanswered until 2003, when Sherman Schultz, a retired optometrist, noted that Lowell's telescopic configuration simulates tools used to examine patients for cataracts. The tiny 0.5 mm aperture is likely to have cast shadows of Lowell's own blood vessels and floaters in the vitreous body of his eye onto his retina, making them visible. Thus Lowell's elaborate maps of Mars and Venus were probably diagrams of the blemishes and structures in his own eyes.

By Emily Parsons-Lord

49. KONSTANTIN TSIOLKOVSKY, WEIGHTLESSNESS

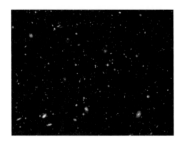

50. HUBBLE ULTRA DEEP FIELD PHOTOGRAPH

51. LAST FUTURIST EXHIBITION, ST. PETERSBURG

52. "WOW!" SIGNAL, BIG EAR OBSERVATORY, OHIO

When Jerry Ehman sat at his kitchen table on August 18, 1977, and saw six numbers and letters on a computer printout in front of him, he took a red pen, circled the letters, and wrote: "Wow!"

At the time, Ehman was teaching at Ohio State University while volunteering for the collaborative search for extraterrestrial intelligence (SETI). Every few days, a messenger biked over from The Big Ear, Ohio State's giant radio telescope, and handed Jerry computer records of radio waves collected from space. Every night, SETI recorded a long list of letters and numbers, one long string for every one of the fifty channels that were scanned by the telescope. The series that caught Ehman's attention, 6EQUJ5, originating in the constellation Sagittarius, grew in strength and then subsided within thirty-seven seconds—the amount of time it takes the Big Ear scanning beam to actually survey a given point in outer space.

The series looked like what SETI researchers expected a signal from an alien civilization to look like. The team tried repeatedly to relocate the signal in the month that followed, but with no luck. Ten years later, new researchers embarked on searches for the "Wow!" signal again and still found nothing. There is still no consensus as to what the "Wow!" signal may have been.

53. CONSTRUCTION OF HOOVER DAM, ARIZONA

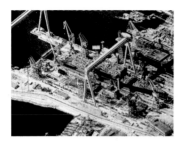

54. NIKOLAEV SHIPYARD, BLACK SEA

Samuel Loring Morison, an American Navy Intelligence Analyst, was convicted of espionage in 1985 for giving the British *Jane's Defence Weekly* three photographs taken by a top-secret KH-11 reconnaissance satellite. Morison was the first American government official to be prosecuted and convicted under the Espionage Act of 1917 for giving classified information to the press.

While working for the navy, specializing in Soviet amphibious and mine-laying vessels, Morison also served as a paid consultant for the American section of *Jane's Fighting Ships*, an annual reference book. He took the three photographs from a coworker's desk, cut the classified control markings from them, and mailed them to an editor at *Jane's*. The images show the general layout of the Nikolaev 444 shipyard in the Black Sea, as well as a Kiev class aircraft carrier, then known as the *Kharkiv*, and a landing ship.

Morison was arrested on October 1, 1984, after a search of his apartment in Crofton, Maryland, revealed several hundred government documents, some of them classified, that he had obtained while he had top security clearance. He was charged with espionage and theft of government property. No real motive was cited in court documents.

Morison was pardoned in 2001 by President Clinton on his last day in office, despite vigorous objection by the Central Intelligence Agency.

55. LEVITTOWN, NEW YORK

56. THE GREAT WAVE OFF KANAGAWA, KATSUSHIKA HOKUSAI

57. SEPARATION WALL, JERUSALEM

On a family trip to Tel Aviv in 2010, Simon Chetrit snuck out of his cousin's bar mitzvah with an overnight bag. He wanted to see for himself the Separation Wall along the West Bank. "I come from a very orthodox Jewish family with corresponding political views," Chetrit says. "I'm also a first-generation Moroccan [American], born and bred in New York City. I wanted to see what normal life in Palestine is like." He found a guy online who was offering his couch for a night, and took a bus later that evening.

The state of Israel began building the twenty-two-foot-high, miles-long wall—which is taller and wider than the Berlin Wall—in 2002. Chetrit spent a day walking along the wall. "It's amazing the way the thing has carved up the land, the whole area," he says. "I saw a gas station along what used to be a road. But now the path is so narrow that cars can't get to it. That's probably why there are so many derelict businesses in its shadow." The thing that made the biggest impression on Chetrit was the graffiti on the wall, especially a white line someone had painted along the wall's entire length.

Chetrit, twenty-three, didn't tell his family that he was planning to see the wall, for fear of repercussions. "I'm not going to say I wasn't outside of my comfort zone," he says. "It was Ramadan. I heard loud explosions, which were just fireworks—but still." He photographs portraits for a living, but is also working on a photo essay comparing images of Tea Party rallies he shot a few years ago with images of Occupy movement protests from a few months ago. "I like to see the realities of both sides of things."

58. ILLINOIS STATE PENITENTIARY AT STATEVILLE, JOLIET, ILLINOIS

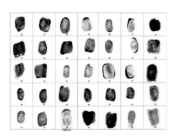

59. FINGERPRINTS

60. INKED HANDPRINT AND CONTRACT

In 1958, Sir William James Herschel, a British magistrate in Nuddea, India, asked a local businessman, Rajyadhar Konai, to dip his palm in ink and press it onto the surface of a legal contract. Herschel had developed the practice years before while working for the East India Company. To formalize agreements with Bengali contractors, Herschel finalized them with a handprinted seal. Although he had little knowledge at the time of the singular qualities of a handprint, he recognized that the action itself lent weight to the exchange.

As a magistrate in Nuddea, Herschel was charged with issuing pensions to locals and preventing fraud. Because many locals were illiterate, capturing signatures was often difficult. Herschel began requiring fingerprints as a form of signature and as a way to identify people.

Although Herschel wasn't the first to recognize the fingerprint's ridges, spirals, and loops, he has since been credited with being the first to employ them to legalize government and business relationships.

61. OPERATION CROSSROADS BAKER, BIKINI ATOLL

62. NGUYEN HUU AN, PHAN TI CUC, AND NGUYEN THI THANH, HUE

Phan Ti Cuc lives with her children, Nguyen Huu An and Nguyen Thi Thanh, in a small house near the town of Hue, Vietnam. She is a single mother: her husband committed suicide by drinking herbicide shortly after Nguyen Thi Thanh's birth. Phan Ti Cuc survives on financial aid from the government and the Vietnamese Red Cross.

Nguyen Huu An and Nguyen Thi Thanh have literally become the poster children for the long-term effects of Agent Orange, a toxic herbicide the U.S. military sprayed over the Vietnamese countryside during the Vietnam War. The chemicals didn't simply kill plants; four hundred thousand people were killed or severely maimed and five hundred thousand children have been born with birth defects.

This picture was taken by Swiss photographer Roland Schmid, who received a 1999 commission by a group of NGOs, including the Swiss Red Cross and Caritas, to document people afflicted by the aftermath of Agent Orange. Vietnamese government officials brought Schmid to see Phan Ti Cuc's family. "They didn't ask the family if they wanted to be interviewed or photographed. They simply had to accept," Schmid recalled.

"I had been visiting hospitals and families for four weeks, with a journalist," says Schmid, who lives in Basel, Switzerland. "It wasn't easy and often very sad. I remember the mother's voice. It was weak."

Schmid won a Swiss Press award for this series of images, which appeared in exhibitions, a book, and newspaper features. "The goal was to enlighten people," he says, "and to raise money for the area."

63. LUDWIG JOSEF JOHANN WITTGENSTEIN'S APPLICATION FOR A BRITISH CERTIFICATE OF NATURALIZATION

Austrian-born philosopher Ludwig Wittgenstein applied for British citizenship in 1938, as Nazi Germany expanded across central Europe. A British passport would have granted him travel to and from Vienna, his hometown, from which he faced exile. This Oath of Allegiance, which he submitted during the naturalization process, includes accounts of his life in Cambridge, vacations, and testimony from character witnesses, such as John Maynard Keynes.

The document also includes commentary from government officials: "This man is presumably sympathetic to Communism, as when he first came to England he stayed with the Communist Maurice DOBB, and in 1935 visited Russia. Apart from this we know nothing to his detriment."

64. JAPANESE INTERNMENT CAMP, HEART MOUNTAIN RELOCATION CENTER, WYOMING

During World War II, the U.S. government removed approximately 120,000 Japanese and Japanese-Americans from their homes on the West Coast and incarcerated them in internment camps for three years. The Heart Mountain Relocation Center, in Heart Mountain, Wyoming, was one of ten such camps. Heart Mountain opened in August 1942; by October, ten thousand prisoners had been relocated to a small barracks, with a shared mess hall, shower, and latrine. Food was scarce, and prisoners were required to work at the camp—and paid less than a fifth of the average American salary for the same position.

65. CAIRO, EGYPT

66. CAPTAIN AMERICA

67. AHMAD SAID MOHAMMED BSEISY, PASSPORT

Ahmad Bseisy fled his ancestral village of al-Maliha after Zionist militias destroyed a nearby community in the 1948 massacre at Deir Yassin. He took up residence in Jerusalem with his wife and children, where they would still be close enough to tend to their farmland by day but avoid the threat of raiders and violence at night. When Jerusalem itself became too violent, Ahmad moved once again, this time to Bethlehem. "And then one day he couldn't go back to the village," explains Ayah Bseisy, Ahmad's granddaughter. "He spent the rest of his life waiting for things to calm down, to return to his land, but things only got worse. Hundreds of families were uprooted and are still waiting for their right to return."

Ayah Bseisy works for an NGO in the West Bank that provides aid in rural Palestinian villages. She wrote a chapter in *People Without a Country: Voices from Palestine*, describing the day the gates were blown off of her school's entrance by men in jeeps, and how she escaped a playground with her brother after their friend was shot to death. Ayah never met her grandparents. Her mother died when she was three years old, her father when she was seventeen. She combs through her family's old papers, looking for her own history.

"Being a refugee is something that can't be described," she says. "You always feel that you don't belong to this life. You feel robbed."

Ayah has posted online photographs and other records that offer bits of information about who she is and where she comes from in an effort to connect with others from the same village. "These documents prove that my family belonged to Maliha," Bseisy says.

68. YVONNE CHEVALLIER

In 1951, Yvonne Chevallier discovered her husband, French Cabinet Minister and war hero Pierre Chevallier, having an affair with the wife of another man. She went to the local police station to apply for a handgun permit (explaining that she needed the weapon for self-protection due to her husband's prominence in government) and then purchased a Mab 7.76mm pistol. On August 12 of that year, while her husband dressed backstage for a speech, Chevallier pleaded with him to end the affair and was harshly rebuffed. She pulled out the pistol and threatened suicide; Pierre told her to wait until he'd left the room. Instead, Chevallier shot her husband five times and then sat patiently by his body for the police to arrive. The ensuing trial was a national sensation, ending with Chevallier's acquittal by seven male jurors.

Chevallier subsequently moved to French New Guinea, where she spent the rest of her life working as a volunteer nurse for the poor.

The image was included in Edward Steichen's 1955 *Family of Man* exhibition and catalog, where it is presented without accompanying text. "I was struck by Steichen's choice to place this image in the area of the exhibition where he used notions such as 'rebels,' 'justice,' and 'public debate,'" explains photo theorist Ariella Azoulay. "In the context of the exhibition, but also in that of women's civil status at that time, it is a courageous choice to frame a woman who was accused of a murder as a claimant. In conditions of oppression of women and their subjugation under the marriage contract, justice requires more than the written law."

69. POLICEMEN, NEPAL

70. BABYLONIAN MATH TABLET YBC 4713

YBC 4713 is an ancient Babylonian mathematical tablet that includes thirty-seven word problems having to do with geometric calculations. In these problems, the area of a figure is given, and the width and the length have to be calculated. Most of the Mesopotamian mathematic tablets known to modern scholars come from the second half of the Old Babylonian period (ca. 1800–1600 B.C.), and most of them are so-called school-tablets. Both writing and mathematics were taught in scribal schools, usually connected to the administrative offices of the palace or temple. Mathematical texts in the Sumerian and Akkadian languages were written in cuneiform script with a reed stylus on clay tablets, from left to right, and column by column.

Old Babylonian mathematics used the powerful sexagesimal place value system for sophisticated and abstract calculations. A thousand years later, the same system was adopted for Babylonian astronomy, and even today we use the system in dividing time into hours, minutes, and seconds.

The following is an example of a problem from the YBC 4713 tablet: "The area equals 10.0. I multiplied the length by a certain number, and got 2.30. I multiplied the width by a certain number, and got 1.20. The number I multiplied by the length exceeds by 1 the number I multiplied by the width. What are the length and the width?"

By Ulla Kasten

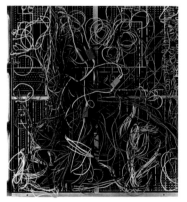

71. EARLY IBM COMPUTER

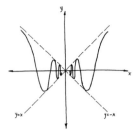

72. MONSTER FUNCTION, RAFAEL NÚÑEZ

"Logic sometimes makes monsters," remarked Henri Poincaré in 1899 about the recent invention of pathological or monster functions in mathematics. Equations such as the one shown here, $f(x) = x \sin(1/x)$, bore no relationship to any reality outside mathematics. They were purely mathematical constructs.

"In former times, when one invented a new function it was for a practical purpose," lamented Poincaré. "Today one invents them purposely to show up defects in the reasoning of our fathers . . ." The monster functions demonstrated that mathematics might not have any special relationship to reality and implied that mathematics might be a purely human construction, subject to the same foibles and idiosyncrasies as other human endeavors.

Cognitive scientist Rafael Núñez, who drew the function shown here, insists that "mathematics has a human face." For Núñez, there is much at stake in understanding mathematics as a human construction rather than something transcendental: "We humans create amazing concepts, and we are responsible for their outcomes." If humans create mathematics, then humans are responsible for what we do with it. To believe that mathematics is an intrinsic feature of the universe is, for Núñez, to believe in dogmas.

"Dogmas generate and sustain arbitrary abusive power . . . by showing how mathematics is a human construction, we can learn about ourselves in a dogma-free manner, and realize how deep, creative, and fruitful our animal mind is!"

73. RORSCHACH TEST

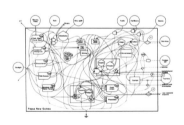

74. CYBERNETIC DIAGRAM OF PAPUA NEW GUINEA, H. T. ODUM

In the late 1950s and the 1960s pioneering ecologist, Howard T. Odum, developed a visual language to represent ecosystems. This cybernetic diagram, Energy Diagram of Papua New Guinea, is one such example. The image juxtaposes widely varying elements (it contains sunlight, as well as assets, people, culture, and religion) in what appears to be a single organizational matrix. The diagram is a prime example of Odum's work in the field of systems ecology, an interdisciplinary approach that applied general systems theories to ecology and biology.

Many in the physical and social sciences in the postwar period thought that ideas about the nature of communication and feedback that developed in systems theory and followed general and universal mechanical principles could be broadly applied to study ecological, mechanical, and social systems. One of Odum's main conceits was the idea that any complex ecological or social system could be modeled like an electrical circuit, that is, represented in terms of flows of energy and information through the system.

By Max Symuleski

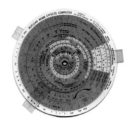

75. NUCLEAR BOMB EFFECTS COMPUTER

76. DIVISION BY ZERO

77. "ONTOLOGICAL PROOF," KURT GÖDEL

*1970, or Gödel's Ontological Proof, is the mathematician's formal argument for the existence of God. Ontological proofs for the existence of God date back to the Middle Ages; the eleventh-century logician Anselm of Canterbury reasoned that: "God, by definition, is that for which no greater can be conceived. God exists in the understanding. If God exists in the understanding, we could imagine Him to be greater by existing in reality. Therefore, God must exist."

Formal proofs such as *1970 are intended to measure the truthfulness of propositions. This practice comes from geometry. Mathematicians like Kurt Gödel carried the practice of proof writing from geometry into more abstract branches, like set theory. Gödel became famous for proving that mathematics cannot be formalized into an inclusive and consistent logical system because the natural numbers have relations that are true but undecidable. *1970 is a proof for God expressed in the language of formal logic.

The basic premise of Gödel's argument recapitulates St. Anselm's: if we can conceive of "something greater than which can be thought," then we must include existence in its set of properties, because existence is presumably greater than non-existence. Gödel's version defines positive properties, deduces that positive properties are necessarily positive and that the set of all positive properties is compatible, assumes existence as a positive property, then concludes that if God possibly exists, God necessarily exists. An arguable weak link of the proof is Axiom 2, the presumption that either a property is positive or its negation is positive.

Gödel presented the piece quietly to friends at the end of his life, not as a grand statement, but as a test upon the limits of his field.

By Laura Grieg

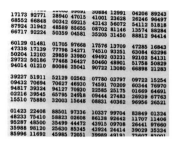

17173 92771		30884 12991	04206 89243
68552 66848	28940 47015	41001 23628	26246 96497
87924 31942	00342 69215	43143 56072	54112 51818
66717 92224	48589 85655	08702 81146	13574 88284
	50359 04581	35205 31450	88812 94414
60129 01481	01705 97668	17576 13709	47285 16843
47338 17139	77798 34271	74510 93251	63084 63296
50204 12103	29859 33980	49492 03231	92102 54131
29722 50186	77468 36427	50460 48901	51758 50829
94014 01210	80086 25041	90722 13080	66898 21283
39227 51911	52129 02563	07780 03797	09722 15254
09432 70694	70627 46930	74591 70209	60344 76970
94817 39324	94127 70930	22585 25175	01669 64691
02216 39545	65795 24918	09444 27483	26939 67763
15510 75880	32003 15648	08831 40362	96956 26521
01423 22408	88501 97336	10337 99704	83849 01324
48233 75410	58823 02608	86138 92642	11707 11406
95287 48500	35499 44472	43915 09708	26441 97660
35988 98120	25630 85345	43924 24414	39029 35334
85996 11692	45985 72051	39989 49191	73607 83001

78. *A MILLION RANDOM DIGITS WITH 100,000 NORMAL DEVIATES*, RAND CORPORATION

The RAND Corporation's *A Million Random Digits with 100,000 Normal Deviates* was a popular scientific reference text whose tables of random numbers have been used in fields from aerospace engineering to poll taking. According to RAND, the book was a response to an overwhelming demand that arose within the corporation for randomly generated series of numbers, "needed to solve problems of various kinds by experimental probability procedures."[1] *A Million Random Digits* was a series of random numbers longer than any other previously published, the production of which involved years of computation and assessment by RAND mathematicians.

The first tables of random numbers were produced with the aid of an electronic roulette wheel, providing what mathematical calculation or computation alone could not: the element of chance. The numbers were then assessed for possible patterns and further randomized until they were found to be sufficiently pattern-free before publication. Although RAND Corporation at that time possessed some of the most sophisticated computing equipment in the world, there was then, and still is, no way to provide true randomness without a mechanical (that is, physical) process. Computers are now able to produce sufficiently long strings of pseudo-random numbers for most real-world applications, but all computational methods necessarily rely on algorithms. Randomness, then, is a problem that points to the very limits of our mathematics.

By Max Symuleski

1. RAND, *A Million Random Digits with 100,000 Normal Deviates*, 2nd ed. (Glencoe, IL: Free Press, 1955; Santa Monica, CA: RAND, 2001), ix.

79. SPACEWALK, ALEXEY LEONOV

80. AIR FRANCE CONCORDE

81. SAND DUNES, CARSON DESERT, NEVADA

82. COMPUTERS ON PARADE, EAST BERLIN

On July 4, 1987, the East German district of Erfurt participated in a Jubilee Procession celebrating Berlin's 750th anniversary and the renovation of the city's major railway station. Residents and workers paraded through the city's streets with desktop computers in tow.

Thomas Uhlemann's photographs of Berlin mark the transitional years just before the Wall fell through German reunification. After reunification, that celebrated rail station would become the connection point for a new train connecting the former East and West Germany. Uhlemann's later images capture West Berlin checkpoint crossings and Bruce Springsteen performing "Tunnel of Love" in East Berlin.

This photograph, like much of Uhlemann's work, belongs to the German National Archive, which in 2009 donated its entire holdings to Wikimedia Commons—one of the largest donations of images to an online repository.

83. SKY SPORT 24, ITALY

84. TOKYO

85. THE BLACK HOLE, LOS ALAMOS, NEW MEXICO

The Los Alamos Sales Company, also known as The Black Hole, is the unofficial scrap yard for the nearby Los Alamos National Laboratory (LANL), one of two labs in the United States that designs nuclear weapons. Los Alamos was built in 1943 as the Manhattan Project's research and development headquarters.

Frustrated with the amount of waste produced by the lab, Ed Grothus (1944–2009) opened The Black Hole for business in 1953 while working as a high-level machinist in the Los Alamos weapons group, in order to repurpose the high-tech nuclear junk that was accumulating. At LANL, Grothus worked on bombs thirty times smaller and more powerful than those dropped during World War II.

During the Vietnam War, Grothus decided he could no longer justify contributing to the lab's nuclear mission, so he left to focus on his salvage business instead. He purchased an old Lutheran church, a neighboring Piggly Wiggly grocery store, and several other buildings, and set up shop full-time.

Ed's daughter Barbara recounts that, "During his time [at LANL] he used the first Marley camera at some point in his work. This camera could take a hundred thousand images per second. But they also soon had cameras that could take a million images per second using mirrors. They used these to find out exactly when the forces in the bomb were at 'critical mass.'" Later in life, Grothus held his own anti-bomb protests, dubbed "critical mass," at The Black Hole's church.

86. WATUSI HIGH EXPLOSIVE EXPERIMENT, NEVADA TEST SITE

87. "BRINGING SCIENCE TO INDIA," UNION CARBIDE ADVERTISEMENT

88. DANDELION

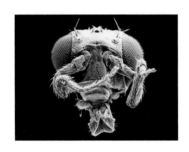

89. GENETICALLY MODIFIED FRUIT FLY WITH LEGS INSTEAD OF ANTENNAE

90. STONE MEDALLION, CARL JUNG, BOLLINGEN, SWITZERLAND

91. CHERRY BLOSSOMS

92. THE PIT SCENE, LASCAUX CAVE

93. GRINNELL GLACIER, GLACIER NATIONAL PARK, MONTANA, 1940

94. GRINNELL GLACIER, GLACIER NATIONAL PARK, MONTANA, 2006

95. NARBONA PANEL, CANYON DE CHELLY, NAVAJO NATION

In 1805, Antonio de Narbona led an expedition of Spanish soldiers, accompanied by allied Native Americans, into Canyon de Chelly in the Navajo Nation to attack the Navajo tribe. When the Navajo learned of Narbona's impending arrival, they scaled the canyon's vertical cliffs, finding refuge in a cave where the Spanish could not reach them. Narbona's men fired upward; bullets ricocheting from the walls of the cave took the lives of all of those inside. The cave is now known as "Massacre Cave." Although the Spanish claimed to have taken the lives of ninety Navajo warriors in addition to twenty-five women and children, the Navajos recall the dead to have been mostly women, children, and the elderly, as the men were away hunting during the Spanish invasion.

The massacre is depicted in this Navajo pictograph in Canyon de Chelly. The image shows the Spanish cavalry wearing flat-brimmed hats and long winter capes. Soldiers on horseback carry muskets, followed by a priest wrapped in elaborate robes.

96. WATERSPOUT, FLORIDA KEYS

97. SUEZ CANAL, EGYPT

98. DUST STORM, STRATFORD, TEXAS

99. PREDATOR DRONE, NORTH WAZIRISTAN

"Witnessing a drone hovering over Waziristan skies is a regular thing," says Noor Behram, who shot this image outside his house in Dande Darpa Khel, North Waziristan.

Since 2008, Behram has been documenting the aftermath of drone attacks in Pakistan's tribal areas, the hub of the Central Intelligence Agency's remote assassination program. When Behram learns of a strike, he races toward ground zero to photograph the scene. "North Waziristan," he explained, "is a big area scattered over hundreds of miles, and some places are harder to reach due to lack of roads and access. At many places I will only be able to reach the scene after six to eight hours." Nonetheless, Behram's photographs are some of the only on-the-ground images of drone attacks.

"[The] few places where I have been able to reach right after the attack were a terrible sight," he explains. "One such place was filled with human body parts lying around and a strong smell of burnt human flesh. Poverty and the meager living standards of inhabitants is another common thing at the attack sites." Behram's photographs tell a different story from official American reports, which consistently deny civilian casualties from drone attacks. "I have come across some horrendous visions where human body parts would be scattered around without distinction, those of children, women, and elderly," he adds.

For Behram, this particular photograph is nothing exceptional. "This was like any other day in Waziristan," he says, "coming out of the house, witnessing a drone in the sky, getting along with our lives till it targets you. That day it was in the morning, and I was at my home playing with my children. I spotted the drone and started filming it with my camera, and then I followed it a bit on a bike."

100. FLOWERY STEPPES BETWEEN BARNAOUL AND TOMSK IN SIBERIA, RUSSIA

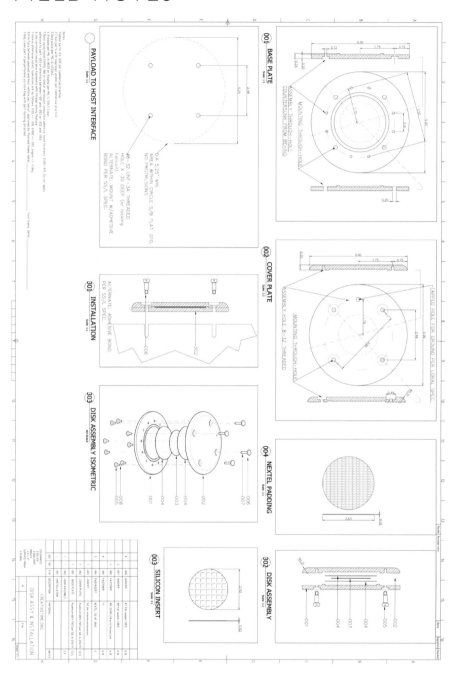

THE ARTIFACT COVER ETCHING

JOEL WEISBERG

Working with Trevor on the Artifact cover offered the opportunity to consider the meaning of *very* deep time in the context of our and others' civilizations. It was fascinating to reflect on how to communicate so far into the future—far longer than the human race has existed, and even longer than the lifetimes of some astronomical phenomena.

We looked especially to the ideas of the Pioneer Plaque creators for guidance, while recognizing some fundamental differences between the two artifacts. For example, although we both want to inform the discoverers of the *time* that these artifacts were created, communicating the *location* of origin is very different in the two cases. The Pioneer Plaque has left the solar system, but the geostationary Artifact is bound to the Earth forever, so there was no need to inform the discoverers of the Artifact's place of origin. In both cases we use scientific concepts that we presume to be universal to communicate these ideas, and in both cases it seems reasonable to assume that the Artifact discoverers will be at least as technologically advanced as we are in order to be able to discover them! It is also fascinating to revisit today the choices made in the 1970s—some have withstood the passage of several decades, whereas others have already been superseded by new scientific developments. It's our hope that a discoverer in the deep future will be able to understand both.

To enable the discoverers to date the origin of the Artifact, we looked for patterns in nature that could survive recognizably for a very long time, and whose changes are sufficiently predictable to enable the discoverers to date the Artifact by comparing their observations of the patterns with ours as recorded on the Artifact. For example, the relative locations and patterns of the "fixed" stars are actually changing slowly over time, and astronomers have cataloged not only their positions but also their motions across the sky. Presumably the discoverers will also have done so—or will be inspired by their discovery to do so. Therefore they could date the Artifact by comparing their star patterns and motions with those on the Artifact. (The stars on the Artifact are depicted as small solid circles.)

As another example, the earth's axis of rotation, whose northern extension into the sky now points at the star Polaris, traces a giant circle through the sky every twenty-six thousand years, so the discoverers could determine the time elapsed since the creation of the Artifact by seeing how far the pole has wandered from its location on the Artifact. (The current southern extension of the earth's spin axis is shown on the Artifact as the intersection of the long horizontal and vertical line.)

A third example is the Earth map. Because our planet's tectonic plates are moving, the discoverers will be able to estimate the time elapsed since construction of the Artifact by comparing their own Earth map with the one on the Artifact cover (Trevor showed political boundaries on the world map as a hint that our world is fractured along more than geologic lines). In this sense, the placement of the current starfield, the current pole, and the current Earth map on the Artifact create "clocks," enabling the discoverers to determine the time elapsed since the Artifact's creation.

> Opposite: Artifact technical drawing by Mason Juday

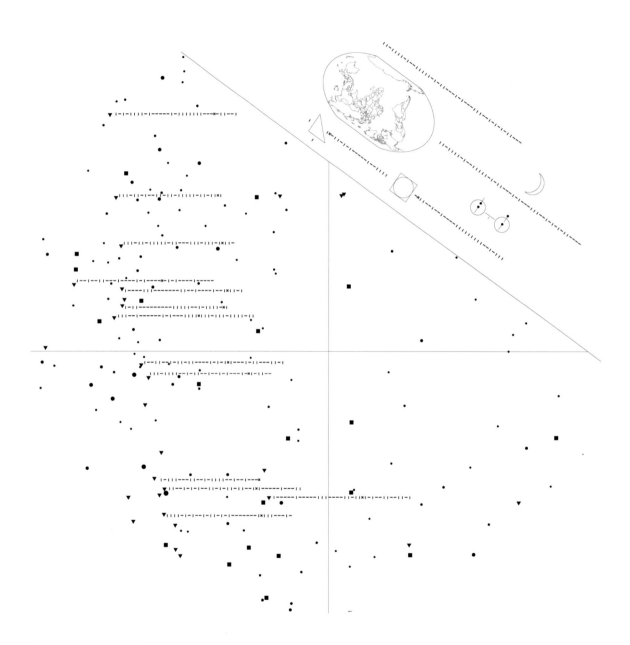

Artifact cover etching

Unfortunately, each of these clocks fails after some period of time. For example, after some thousands of years, many of the stars shown on the Artifact will have drifted too far from their current position to be recognized; others, currently not shown, will have drifted into the pattern; and over long enough timescales some will die and others will be born. (Stars live for millions or billions of years.) Similarly, the location of the earth's pole at the discovery of the Artifact is a fine clock for twenty thousand or so years. However, not long after that the circular motion of the pole will repeat, and the size of the circle made by the pole will slowly change by a few degrees. So although discoverers might be able to determine how many cycles have elapsed, on longer timescales it is possible that there will be chaotic and unpredictable changes, in which case this clock will become irreparably broken. Finally, it is difficult, at least with our current level of knowledge, to predict the motion of tectonic plates over long timescales with much certainty.

We needed a less transient and better behaved set of reference points and clocks to set for discoverers finding the Artifact far in the future. The squares on the star map represent the brightest cosmic point sources of radio waves, most of which are the cores of distant galaxies. These objects are so far away that they will move very little, if at all, over very long periods of time, and they are expected to give strong radio emissions for tens or hundreds of millions of years. If the Artifact is recovered within these timescales, the discoverers will surely be able to match the Artifact pattern with their observations. Even over longer times, they will still find quiescent galaxies at these locations. Therefore these radio sources seem to be an ideal means of providing a reference against which all other celestial objects can be measured, both now and at the time of discovery.

We also wanted to express time intervals in some unit that could be understood by any civilization as "advanced" as ours (because they will not know, for example, what time period a *second* is.) Like the Pioneer Plaque designers, we chose as our basic interval the time between crests of a radio wave emitted by the hydrogen atom when it changes between two fundamental states. The two linked circles near the crescent moon represent the hydrogen atom in the initial and final states of this "hyperfine transition." Our drawing is identical to that carried on the Pioneer Plaque. All time intervals (except one noted shortly) are expressed in units of this atomic hydrogen radio wave time interval. We use base-2 or binary notation to represent numbers of this time interval, with vertical lines representing the numeral 1, horizontal lines representing the numeral 0, and Xs separating the whole and fractional parts (as in base-10). Base-2 arithmetic seems a good choice because it is the most basic; it is used, for example, by almost all digital computers. The base-2 numbers near the circle inscribed in the square and the triangle represent various universal geometrical ratios associated with the drawings, which should enable the discoverers to verify that they understand *our* math and *our* numerical symbols.

The binary number near the earth's south pole gives the current time interval it takes for the earth to spin once on its axis with respect to the stars. The discoverers are expected to compare this number with the measured length of day in their time. Because the length of a day on Earth slowly changes, they may be able to date the Artifact on the basis of the difference between these two numbers. (The earth is slowly spinning down and transferring its lost spin to the moon's orbit, thereby enlarging the moon's orbit and the time needed for it to complete that orbit.)

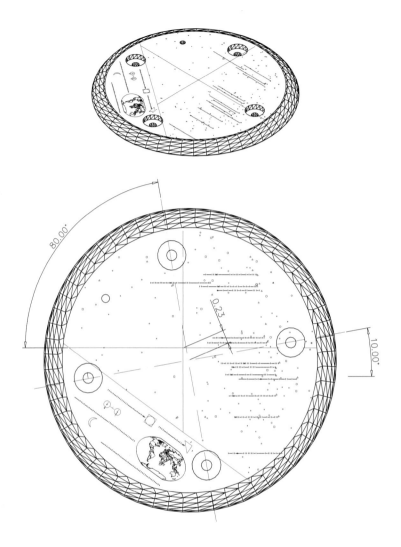

The binary number near the crescent moon gives the length of time needed for the moon to complete one orbit. (This is the only time interval not expressed in units of the hydrogen atom radio wave period, because we would have had to place twenty-nine zeros before the fractional part of the number, wasting valuable space. Instead, it is expressed in units of the earth's rotation period.) Again, the discoverers can be expected to compare this number with their own measurement of the

Time map on Artifact cover

same event, and they may be able to date the creation of the Artifact on this basis.

Finally, the most accurate dating of the Artifact will result from measurements of pulsars. Pulsars are collapsed stars that spin at rates of up to about seven hundred times per *second*, with a beam of radio waves that sweeps the sky once per rotation, much like a lighthouse beam. Each sweep past Earth causes a pulse of radio waves that can be detected on Earth. Pulsars were discovered in 1967, and the Pioneer Plaque designers used them as both time and location markers. However, developments since then enable us to use pulsars as far more accurate time markers than could be done in the *Pioneer* era.

The pulsars known in the 1970s were thought to live for perhaps ten million years. However, in the 1980s, much faster spinning *millisecond* pulsars were discovered. They are expected to live for billions of years and to slow down at particularly smooth and predictable rates, making them the ideal clocks for dating the Artifact far into the future. The labeled triangles represent a mixture of normal and millisecond pulsars, with some of the choices being dictated by their physical location on the Artifact.

Triangles mark the current location of some pulsars, and the base-2 numbers near them represent the time between their pulses. Pulsars' spins are slowing down, mostly in a very regular fashion. Therefore, much as is the case for several of our other clocks, the discoverers could date the creation of the Artifact by comparing our reported times between pulses with theirs and their measurements of the rate of slowdown.

Finally, it was exciting for us to realize that the information inscribed on the Artifact will also serve as a true scientific trove for the discoverers. Every quantity that we relay to them represents *today*'s values for various physical phenomena. They will be able to compare these values with the values that they measure at the time of discovery and then see how well their scientific theories predict the changes. It will be like unearthing a scientific time capsule from the distant past. We can expect that they will be able to test and improve some of their theories as a result of these analyses. They will also be able to assess the state of our technology by noting the types of data that we have placed on the Artifact and the types that we have not included because we have not yet discovered them!

TALKING MATHEMATICS TO ALIENS?
(GET REAL! . . . OR HAVE FUN WITH ANTHROPOMORPHISM 101!)

RAFAEL NÚÑEZ

Breaking news! Dr. Ellie Arroway, a competent scientist, has just gathered conclusive radio-transmission proof of intelligent aliens—and in a matter of hours, the attractive and passionate scholar finds herself talking with high-level politicians about the urgency of understanding those extraterrestrial messages. Addressing a respected lawmaker's mocking question, "Why don't the aliens just speak English?" she serenely—and without the slightest hesitation—replies: "Mathematics is the only truly universal language, Senator."

Wow! What a hero! What a deeply reassuring, breathtaking line! We all want to believe this is true. (Please, pretty please, don't tell me it is not true!)

The scene is from the movie *Contact,* written by the Pulitzer prize–winning author Carl Sagan, in which the heroic Ellie (played by Jodie Foster) instantiates a commonly held belief that mathematics, as we know it on Earth, is "out there," constituting the very fabric of the cosmos and transcending the existence not only of human beings but also of any possible beings of any kind in the entire universe. This view is unconsciously accepted by millions, nurtured by other such movies and popular books (some of them written by well-known scientists, like Carl Sagan himself) that deliver the same message. Furthermore, this view of mathematics is not confined to Hollywood and popular science books; it is also alive in many academic institutions that endorse different forms of Platonism. The belief is deep and often unnoticed and unquestioned: mathematics, whose existence transcends human beings, is indeed the ultimate universal language.

But is it? In fact, this belief is no more true than the claim that E.T. is a cute creature—it is a human belief based on human impressions and evaluations. In our book *Where Mathematics Comes From*, George Lakoff and I called the view that supports this belief The Romance of Mathematics. Despite the fact that many smart and serious scientists, mathematicians, and philosophers vigorously defend such belief in the name of science, it is simply that: a belief. And it is scientifically untenable. Here is why.

First, science has a paramount commitment to empirical observations. Findings in contemporary cognitive science—the multidisciplinary study of the mind—show that mathematics is about thought, language, and cognitive inferential mechanisms, and these in turn are sustained by the biological processes and structures that make them possible. Because no actual forms of extraterrestrial aliens (simply *aliens* from here on)—dead or alive—have ever been documented empirically, such beings are, scientifically, nonentities. Very much like angels, phantoms, or Donald Duck, aliens, as we know them, are the product of human imagination; therefore no scientific proposals should be made on their behalf.

(How serious would it be to propose a theory that, consistent with evolution, biologically explains why Donald Duck doesn't age?)

Second, if mathematics is the only truly universal language—a language that aliens are presumed to speak fluently—which mathematics is it? There is not just one mathematics, but many, of which many are internally consistent but mutually inconsistent with each other. Some set theories with certain axioms hold certain facts (theorems) to be true, whereas others with different axioms don't. In some geometries, parallel lines do not intersect; in others, they do. Some forms of analysis postulate (and need) the existence of infinitesimal numbers, while others reject them. And so on. Which one of these multiple possible mathematics is the "truly universal language" is a matter of historical timing, taste, and familiarity with the various mathematical forms and techniques. If your alien is like a clever Babylonian of 2000 B.C. and you speak in the transfinite arithmetic of the nineteenth century, she won't understand a word you are saying. Not a word!

In conclusion: angels, phantoms, Donald Ducks, and aliens all are heavily anthropomorphized theoretical creatures, which by virtue of their very properties cannot, in any serious way, inform science as we know it. Of course, there is nothing wrong per se with believing that mathematics is the only truly universal language, allowing us to talk to aliens (if they actually do show up!). But let's not fool ourselves into thinking that if we do so, we are doing it in the name of science. If we want to believe that talking mathematics to aliens makes sense, we must humbly accept that we are anthropomorphizing, big time. And then we are free to write novels, make films, and have fun!

BRIAN L. WARDLE AND KARL BERGGREN

The choice of materials for any object destined for space is usually undertaken through a process called *materials selection*, which focuses on maximizing or minimizing engineering properties such as strength, stiffness, electrical conductivity, and density (mass). In aerospace we are most often concerned with mass–specific stiffness and strength—we want materials to be strong and stiff, but also light. To build Trevor's time capsule, we were thinking about information density; we had lots of information to put in a tiny form that had to resist the processes of time. The familiar expression "the sands of time" took on a whole new meaning for us, as we considered the unique challenge of thinking far beyond the timescales of any known engineered structures—thinking billions of years into the future.

Because time, not sand, would be working against us, we thought about material diffusion as the real driver of the design. Diffusion is a process in which a structure's shape changes slowly over time through the random motion of its atoms and molecules. As atoms in the material move around and bump into each other, like people at a crowded concert, the edges of the structure—like the boundaries of the crowd—can become ill-defined. If diffusion is extensive enough or goes on long enough, the features in the material can become completely dispersed. (In our analogy, the crowd will disappear!) So, as a materials selection exercise, the team narrowed the major issues down to information density and diffusion resistance. If there are two or more molecule types, then there is a driving force for diffusion, and the clock starts on losing the information. An advantageous design would therefore not mix materials

(like ink on paper), but rather have features carved in a single material.

Once our parameters were set, the excitement took hold, and we engineering professors and Trevor quickly grew to understand one another in a blaze of swapping ideas. We had some far-out discussion: about using 3D nanopatterning of materials to get really high-density information, or using 3D to do color coding. We also talked about mechanisms for explaining—to whatever or whoever found this thing in a hundred million years—how to decode it. Everyone brought a distinct point of view to these discussions, which were quickly followed by the usual reality check of a real engineering project with a schedule and other constraints.

Spacecraft engineering, especially manned spaceflight, places a very high premium on staying the same. Anything that has been proven spaceworthy, especially on multiple successful missions, should be used again. This leads to a somewhat strange conservatism in the aerospace industry. An industry that has done the lion's share of materials innovations in the past uses old, but proven, materials and concepts. The same goes for designs—the 2010-era Crew Exploration Vehicle (CEV) closely resembles a larger *Apollo* capsule from the 1960s, even to insiders.

Because the Artifact is hitching a ride on another's payload, we began to narrow down our choices to materials that have done well in space over the short span of years in which humans have been sending craft into space (we have little more than half a century of experience, which is not very much when you are thinking millions of years into the future).

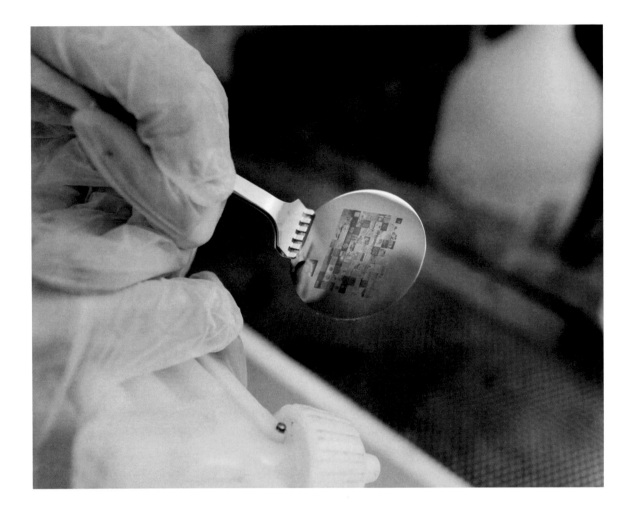

Nano-etched silicon image-disc wafer

Our final selection? The rock we chose to carve Trevor's Artifact images into is the tried and true silicon (or Si), the material of choice for most people working in nano and microfabrication. It is the substrate, or the foundation, on which microelectronics and microelectromechanical systems (MEMS, that is, micro-scale machines) are built. Silicon is considered spaceworthy, as we know a great deal about how it responds to the environment of space, from its use in solar cells, electronics, and telescope elements. It has been stud-ied extensively for ionizing radiation effects, and there are many techniques for making it "rad hard"; that is, resistant to the ionizing radiation effects found in the space environment that the Last Pictures Artifact will be subjected to. (We should mention that the Artifact was eventually packaged inside a gold-plated aluminum clamshell, and other measures were taken to protect it from harmful effects of the space environment .) One of the best things about working in Si—which in this case refers to both single-crystal (SC Si) and silicon oxide—is that there are solid standard techniques for "writing" features into this hard ceramic, using chemical and other means. Features are micron size (one-thousandth of a millimeter) and can be carved into the SC Si substrate. Creating features directly in the SC Si also addresses the diffusion concern, as it avoids introducing any other materials into the equation.

THE ECHOSTAR XVI MISSION

ECHOSTAR CORPORATION

EchoStar XVI will launch in 2012, continuing our tradition of satellite innovation. The bird, weighing more than 6,600 kilograms (14,550 pounds), will be launched into geostationary orbit on an ILS Proton Breeze M rocket carrying 32 Ku-band transponders for delivering direct-broadcast satellite (DBS) signals back to earth. It was built by Space Systems/Loral on the space-proven 1300 platform and is anticipated to have a service lifetime of fifteen years.

EchoStar XVI will replace *EchoStar III*, which is nearing the end of its useful life, and will supplement the DBS video signal delivery of *EchoStar XIII*. The new satellite will ensure continued full use of the direct broadcast spectrum for that orbital slot. When *EchoStar XVI* reaches its geostationary orbital slot at 22,500 miles (36,210 kilometers) above the equator, EchoStar Satellite Services will have twelve owned or leased satellites in orbit.

This state-of-the-art satellite will operate in the 12.2- to 12.7-GHz band, using a single broadcast beam that covers the United States and Puerto Rico and seventy-one spot beams that deliver television signals to specific communities. *EchoStar XVI* will help DISH network satellite provider deliver more high-definition programming, including local HD television broadcast stations, for its more than fourteen million direct-to-home television subscribers in the United States.

EchoStar has always pushed the boundaries of technological creativity. The company is currently in the vanguard of innovators that are forging new electronic pathways to deliver information to an ever-increasing number of consumer devices. It is a natural extension of that mission to provide a new platform—a space-traveling canvas.

EPILOGUE: DEEP FUTURES

"Prediction is very difficult, especially about the future," said physicist Niels Bohr. But it's unlikely that anyone will ever find *The Last Pictures*.

Some have pointed out that the Artifact may be found by humans a few generations from now. One NASA scientist explained to us that some day, in about a hundred years or so, humans will need to clean up the Clarke Belt. At that point, he said, *EchoStar XVI* and the Artifact will probably be brought back to Earth. I disagree. It may be that the geostationary orbit (GSO) will be swept up one day, but bringing communications satellites back to Earth would be pointless and exceptionally inefficient. If, indeed, the world's governments begin an unimaginably expensive program to sweep up space, geosynchronous debris and derelict spacecraft will surely be directed toward "graveyard" orbits a few hundred kilometers above GSO, an zone that's currently considered useless for human purposes. In fact, *EchoStar XVI*'s mission profile already includes an end-of-life maneuver into that spacecraft graveyard just beyond the Clarke Belt.

In the medium term of human time, it's unclear what the future of spaceflight may hold. Subsequent generations may conclude that spaceflight is simply too resource-intensive to be sustainable. They may look back on our rockets and satellites as the eccentric relics of an earlier age's collective psychosis, eyeing them with the same combination of wonder and puzzlement with which we ponder the statues of Easter Island. No, spaceflight is not necessarily here to stay. We tend to think of ourselves as the culmination of an inevitable historical progression, from hunter-gathering to genetic engineering, and from Babylonian stone tablets to Google, with rockets and satellites as a milestone on some great path of progress. But human history is not a grand and unified march towards an inevitable and glorious future. "[W]hat we call 'progress' . . ." wrote the French revolutionary August Blanqui from his prison cell at the Fort du Taureau during the Paris Commune, "is confined to each particular world, and vanishes with it. Always and everywhere in the terrestrial arena, the same drama, the same setting, on the same narrow stage—a noisy humanity infatuated with its own grandeur, believing itself to be the universe and living in its prison as though in some immense realm, only to founder at an early date along with its globe, which has borne with deepest disdain, the burden of human arrogance."

One lesson we can take from Ancient Egypt, the Roman Empire, the early Easter Islanders, the Mayans, and the Anasazi is that human histories have no clear direction. In the long run, histories and civilizations advance and recede like waves lapping on a beach.

Without question, human civilizations will not last forever. Sooner or later, *Homo sapiens* will join the ranks of the vast majority of species that have inhabited the earth are now extinct. The average lifetime of a species is about five million years. For mammal species such as ours, tending to be higher up the food chain and therefore more vulnerable to environmental change, the average lifetime is about one million years. (Humans may be an exception, owing to their ability to thrive in a much larger range of habitats than other mammals.) There is, however, a big asterisk next to these numbers: if we look at the average rate of extinction over the last ten thousand years, it's clear that we are now in the middle of an extinction event more dramatic than the one that killed the dinosaurs nearly seventy million years ago. E. O. Wilson explains that the current extinction rate is "somewhere between one thousand and ten thousand times the rate before human beings began to exert a significant pressure on the environment." At the current rate, "it is safe to say that at least a fifth of the species of plants and animals would be gone or committed to early extinction by 2030, and half by the end of the century." This is not a fringe theory, but mainstream scientific opinion. In response to a 1998 survey by the American Museum of Natural History, seven out of ten biologists agreed that "we are in the midst of a mass extinction of living things" and that "this mass extinction is the fastest in Earth's 4.5-billion-year history . . . this loss of species will pose a major threat to human existence in the next century."

Will another species ever arise on Earth with the combination of intelligence, technology, and imagination to discover the Clarke Belt's ancient machines and set forth to inspect them? There is plenty of time for such a thing to happen—the Artifact will remain in orbit for nearly ten times longer than the amount of time that has passed since animals first walked on land. Nonetheless, it's exceptionally unlikely that another spacefaring species will arise on Earth. Consider how staggeringly unique contemporary human civilizations are in Earth history. An "intelligent" civilization capable of spaceflight could have arisen at any point in the last billion years of life

on Earth. As far as we know, it's happened only once, and only for the briefest moment of human history. Even if a future life form physiologically capable of developing spaceflight emerged, there's no reason to assume that they might take to the stars. *Homo sapiens* has existed for nearly two hundred thousand years. If we discard the idea that history inevitably culminates in civilizations like our own, then our conclusion must be this: our fifty-plus-year experiment with spaceflight is an incredibly rare anomaly, even for a species with the biological and intellectual ability to go to space. Indeed, as Stephen J. Gould put it, "[P]erhaps we are, whatever our glories and accomplishments, a momentary cosmic accident that would never arise again if the tree of life could be replanted from seed and regrown under similar conditions."

The time scales are vast, stretching even the notion of deep time. But some events in the far distant future seem assured. Over the next few billion years, the sun's luminosity will slowly increase and the planet will get much hotter. Eventually the earth's oceans will boil off, and our atmosphere will evaporate, rendering the planet inhospitable to life. Even then, *The Last Pictures* will continue slowly circling the earth.

Beyond that is anyone's guess. In four or five billion years, the deep time of the earth will give way to the cosmic time of the stars, and the sun will expand to become a red giant. No one knows whether, at that point, it will swallow our planet in flames. If it does, then *The Last Pictures* too will be consumed. But if the earth survives, then *The Last Pictures* may continue to quietly haunt the planet as the sun collapses into a white dwarf in about 7.5 billion years. All the while, a slow barrage of micrometeorites will gradually grind away at *EchoStar XVI's* hull.

In a hundred trillion years, star formation will begin to cease throughout the universe. One by one, the points of light that dot the cosmos will go black until they are all gone. By this time a wayward neutron star or pulsar will have come perilously close to what was once the sun, sending a great wave through spacetime, spinning *EchoStar XVI* and *The Last Pictures* off into never-ending darkness.

ACKNOWLEDGMENTS

Creative projects, whether they're books or art works, are rarely the result of a single person's effort. Every author or artist has a group of interlocutors, colleagues, friends, and family, all of whom play a role in shaping a particular work.

This project is different. There is no group of people behind this project. Instead, there is a veritable armada.

Scores of people helped develop the ideas, themes, and formal decisions that went into the image collection and narrative for *The Last Pictures*. All of the following people engaged in sustained conversations about how to develop the project, contributing creative ideas and critical feedback: Ayreen Anastas, Javier Anguera, Ariella Azoulay, Yves Behar, Claire Bishop, Julia Bryan-Wilson, Lowrey Burgess, Ignacio Chapella, Catherine Cole, Mike Davis, Marco Deseris, James Faris, Kate Fowle, Rene Gabri, Aaron Gach, Peter Galison, Renee Green, Craig Gilmore, Ruthie Gilmore, Orit Halpern, Stefan Helmreich, Michele Jubin, Steve Kurtz, Siddhartha Lokanandi, Joe Masco, Yates McKee, John Menick, Lize Mogel, Naeem Mohaiemen, Rafael Núñez, Susan Oyama, Rich Pell, Andrew Pickering, Praba Pilar, Rick Prelinger, Hanna Rose Shell, Philip Rosenberg, Jon Roy, Rebecca Solnit, Sundar Sarukkai, Megan Shaw-Prelinger, Seth Shostak, Pete Skafish, Lucia Sommer, Joel Weisberg, and Eyal Weizman.

An outstanding content team spent about six months researching ideas, collecting images, and engaging in weekly seminars. To go in depth into the full extent of their work would fill several volumes. Listed here is only a quick sampling of their topics. Katie Detwiler researched mutant life forms, monsters, images of the state, passports, fingerprints, alchemical texts, nuclear weapons, and various representations of "the human." Laura Grieg worked on mathematical proofs of God, bizarre engineering projects, renaissance painting, and color theory, and helped build the databases we used to organize our work. Emily Parsons-Lord combed through nearly every archive on the East Coast and then some, tracking down obscure political cartoons, cybernetic drawings, DIY cloning kits, eschatological charts, death certificates, and early agribusiness advertisements. Max Symuleski worked on the history of universal languages, cybernetics, unlikely maps and charts, early computing, vision technologies, early-twentieth-century infrastructure projects, and cat pianos. Anya Ventura worked full-time on "what capitalism looks like," biotechnology, ecology, fruit flies with legs coming out of their faces, and dinosaur footprints. My brother, Jack Paglen, spent several weekends going over each image with me and bringing his exceptional cinematic eye to the montage. He gave critical feedback on the overall tone and made a huge contribution to ensuring that the images work as a sort of silent film.

In the early phases of development, João Ribas, curator at the MIT List Visual Arts Center, told me that "this project won't happen without our help"; he proceeded to set up a fantastically fruitful set of collaborations at MIT. Professor Brian Wardle was always available for late-night emergency advice on one topic or another related to aerospace engineering. Wardle brought Professor Karl Berggren, a quantum nanostructures expert, on board, and together they figured out how to make images archival for billions of years ("Well, no one really knows what happens in *billions* of years," smiled Berggren at one point, "but this should at least work for a couple of hundred million.") MIT graduate student Adam McCaughan dropped his other projects at the end of the semester to implement Wardle and Berggren's designs, spending several eighteen-hour days in an MIT clean room actually fabricating the ultra-archival discs that hold the images. Mark Linga at the List Visual Arts Center was an outstanding host, doing the logistics for my visits, setting up interviews, and managing the project on the MIT side. Executive Director of Arts Initiatives Leila W. Kinney was a crucial advocate for the project. João was right—there is no way this project would have happened without their help.

Although we designed the Artifact to the most conservative specifications we could find, the task of designing, fabricating, and flight-testing something for space travel is not a simple undertaking. Throughout the process, aerospace engineer Josh Levine at Lockheed Martin Space Systems was eminently available to guide the design decisions and offer invaluable technical advice and suggestions. Artist/engineer Marko Peljhan was intimately involved in the project from its outset, offering critical feedback and assistance on both the technical and the artistic side. He is one of my favorite artists, and I'm honored to count him as a friend and colleague. When we started getting into crunch time, former NASA engineer Russ Bardos connected us with Steve Overton at Aerojet, who produced some initial drawings and gave us guidance about the fabrication process. Lee Johnson consulted on space materials, and Sam Nackman turned around a set of preliminary AutoCAD drawings in a single night. Designer Mason Juday worked nights, weekends, holidays, and even on his wedding anniversary, on absurdly tight deadlines, doing numerous drawings and figuring out exactly how the Artifact would come together. I can confidently say that the project wouldn't have happened without Mason's bulldog commitment. David Epner of Brooklyn's Epner Technology stepped up to the plate, turning around a large fabrication job in a matter of days and skipping the winter holidays to make the project happen. With decades of aerospace experience on projects like the James Webb space telescope, he was an invaluable advisor. His gold plating is literally the gold standard for the industry. Finally, when we had only a few days to conduct a series of launch and space simulation tests that usually take more than a month, the team at National Technical Systems in Boxborough, Massachusetts kept their facilities running 24/7 to meet our deadline, and they produced a technical report in a single day. Huge thanks to Shane Boivin, Jason Crete, Maria Drury, Jeff Henn, and Pete McDermott for making the impossible happen. I'm truly humbled by the number of people who took a can-do attitude in the face of insane deadlines and ridiculous production schedules to make *The Last Pictures* a reality.

The team at Creative Time was no less dedicated. Shane Brennan and Kevin Stanton were pillars of reliability, arranging interviews, site visits, and photography sessions across the country. Jessica Schaefer was instrumental in developing a marketing plan around the moving target of a rocket launch, and Alyssa Nitchun helped locate the real fuel of every large project: its funding. Katie Hollander expertly made the bottom lines work, ensured that contracts were in order, and made

it possible to hire additional help on very short notice. I don't know how she made the funding side of this project work, but she did. Cara Starke joined the team while we were in full swing. In a matter of days, she'd learned enough about the technical side to help begin sourcing materials and designers and to drive fabrication deadlines. She consistently and competently fielded plans and managed a huge number of moving parts. Aerospace engineering is easy compared to what she's done.

Philanthropists Mike Wilkins and Sheila Duignan have been supporters of *The Last Pictures* since I began work on it, and they provided substantial funding at a critical moment when we were racking up huge bills with no way to pay them. Mike and Sheila are more than supporters; they are interlocutors and friends. Sheila kindly spent time with me in London, visiting the Old Operating Theater and convincing the staff at the National Maritime Museum in Greenwich that they should ignore their own no-photos policy for the good of my project. Mike Wilkins, an accomplished artist himself, and I have enjoyed many years of conversations about this and other projects. Mike encouraged me to keep working at a time where I'd gotten completely frustrated and wanted to trash the whole thing. I'm eternally grateful for his inspiration.

Lawrence Benenson provided another pillar of support. When I first met Lawrence, I could tell immediately that he was deeply sincere, unjaded, and that his heart was in the right place. We have since become good friends. At several points during the development process, Lawrence provided a much-needed fresh set of eyes. Looking at an early draft of the images, he asked "Where are the people?" His feedback and comments helped me break out of a particularly rigid pictorial space and were helpful toward opening up the final selection. His support was like manna from heaven at exactly the right moment to make *The Last Pictures* happen.

Behind each and every one of my projects is a group of people who spend their days figuring out how to sustain my practice so that I can spend time thinking about the stars. I'm truly blessed to work with such fantastic people: Janelle Reiring, Helene Winer, Allison Card, Manuela Mozo, Tom Heman, Alexander Ferrando, and Michael Plunkett at Metro Pictures Gallery in New York; Claudia Altman-Siegel, Dave Berezin, Daelyn Short-Farnham, and McIntyre Parker at Altman-Siegel Gallery in San Francisco; Thomas Zander, Natalie Gaida, Frauke Breede, Christina Mey, and Kristina Keil at Galerie Thomas Zander in Cologne.

They are the backbone of my work and life. I get emotional just thinking about how generous and committed they've all been and how grateful I feel to be a part of their extended families.

The book you're holding is the result of another hardworking and talented group of people who decided to put in long hours for little reward simply because they wanted the book to happen. Book manager Tiffany Hu kept on top of the enormously complicated logistics and delivery deadlines. Sharmila Venkatasubban spent months working at breakneck speed to track down the people and stories behind the photographs and drafted many of the image texts. Lizzie Hurst and Lauren Altman had the unenviable task of securing rights to use all of the images in a short period of time. They put in a Herculean effort. My literary agent, Ted Weinstein, as always, provided amazingly helpful guidance on how to turn the project into a book. Over the many years we've worked together, Ted has never failed to dramatically improve every project and idea he lays his hands on. Rebecca Solnit graciously introduced me to Niels Hooper at the University of California Press. I'd been a longtime fan of Niels' work, and I'm absolutely delighted to be working with him. Niels brought a can-do attitude to a project that's flirted with the impossible since day one. Instead of being frightened off, Niels took on our tight schedule and complicated set of production needs, seeing them as an opportunity to do something marvelous. I thank him for bringing his vision, dedication, and hard work to this book. Designer Lia Tjandra brought her exceptional combination of magical design sense and hard-nosed work ethic to make an extraordinary looking volume.

Alexis Lowry has been at *The Last Pictures*' whirlwind center from one phase to the next. Despite her "day job" as an art history PhD student at New York University, Alexis has been behind nearly everything that had to happen in order for our Artifact to end up in orbit. When she was first hired, she dug into a European Space Agency catalog of every spacecraft in geostationary orbit and made a gigantic list of every satellite in the Clarke Belt, noting the names of owners, operators, associated launch companies, and vehicles. Then she sat at a desk for two months calling every single company on that list, pitching the project idea to anyone who'd listen. Along the way, she figured out the technical requirements for the Artifact, becoming intimately familiar with topics like "coefficients of thermal expansion," "radiated emission requirements," and "thermal vacuum profiles." After spending Christmas on her cell phone sourcing space-proven fabrication materials and arranging launch-simulation tests, Alexis had become so comfortable with this project's utterly unreasonable timelines and workloads that she came back to work on the book. No doubt her future as an art historian and curator is bright, but if she needs a backup plan, she's become a damn good rocket scientist.

Creative Time Chief Curator Nato Thompson's name really should be next to mine as coauthor of this project. We have been close friends for nearly twenty years as well as nonstop interlocutors and longtime collaborators. The idea for this project emerged from more than a year of long conversations with Nato, beginning in 2007. Nato's efforts turned *The Last Pictures* from a gee-whiz idea into an active in-production endeavor, with a staff, schedule, and budget. Nato's name is not on the cover of this book, but he shares the intellectual and creative credit for all the ideas and thinking that went into *The Last Pictures*. After all, we have often been accused of sharing the same brain—something that I'm not inclined to dispute, because I think Nato is a genius.

I am in constant awe of Creative Time President and Artistic Director Anne Pasternak. She's quite possibly the hardest-working, most dedicated person I know. I can't count the number of times when I've fired off a project-related email to her before going to sleep at 2:00 a.m. on a Saturday night, only to see the response come back with a solution to whatever problem I was having. With her personal warmth and calm voice, Anne has talked me down from many emotional and creative ledges. She has been consistently present and available. What's more, she's responsible for finding our satellite. After months of haranguing telecommunications companies, Anne developed an alternate approach. A frequent public speaker, she took to asking her audiences whether anyone happened to own a satellite company. Eventually, Christophe Nicolas said yes and introduced us to Chase Ergen, who became a key advocate and, in turn, connected us to EchoStar Corporation.

I can't thank EchoStar Corporation and Dish Network enough for making this dream of a project come true. EchoStar President Anders Johnson has been personally involved in every step of the process, and his guidance and stewardship have been consistent and unwavering. At Space Systems/Loral, the manufacturers of EchoStar XVI, I owe my deepest gratitude to Zia Oboodiyat, executive director for EchoStar Programs. Zia was extremely generous and kind with his time, knowledge, and leadership. He was excited and enthusiastic about helping to make the project happen. Over time, I learned that Zia is an accomplished poet and author, and a kindred spirit. My studio wall is now decorated with a signed poem he gave me.

Above all, two people are most directly responsible for the realization of this project. Chris Ergen took up the idea for *The Last Pictures* and became its champion at EchoStar, convincing the company to play host to our peculiar artwork. Chris and I have been able to meet a few times, and I've been astonished not only by the depth of his knowledge but also by his ability to put it toward creative use. No doubt Chris will be one of the key people shaping the future of Earth's communications infrastructure. And although I have never met Dish Network's cofounder, president, and CEO Charlie Ergen, I have the greatest appreciation and admiration for his decision to host the project on his company's spacecraft. He is a true visionary.

Finally, Kate Fowle lived with me and this project through every up, down, and wild side-to-side turn. She has been an almost daily interlocutor and has tolerated hundreds of evenings of my obsessively worrying about the nuances of interspecies communication, or the long-term fate of neutron stars, usually to the detriment of our social life. Kate has saved me from making a great number of truly bad artistic decisions, and her steady ability to think through bureaucracies and to see issues from multiple vantage points has helped smooth out this project's bumpy road on a regular basis. I owe her the world. It doesn't hurt that I'm madly in love with her.

PROJECT SUPPORTERS

The Last Pictures was made possible through the visionary support of:

Lawrence B. Benenson

Sheila Duignan and Mike Wilkins

EchoStar Corporation

Epner Technology, Inc.

MIT Visiting Artists Program

National Endowment for the Arts

Creative Time's major programming support for 2012—2013 has been provided through the generosity of Bloomberg Philanthropies, Ford Foundation, and Lambent Foundation. Creative Time is pleased to acknowledge public funding from the New York City Department of Cultural Affairs, in partnership with the City Council; and the New York State Council on the Arts, with the support of Governor Andrew Cuomo and the New York State legislature.

CREDITS FOR THE LAST PICTURES

Images 6, 10, 11, 13, 25, 26, 44, 48, 59, 73, 75, 75, 76, 77, 78, 85, 95 © 2012 Trevor Paglen.

1. Courtesy of the Israel Museum.
2. NASA/Carla Cioffi.
3. National Oceanic and Atmospheric Administration/Department of Commerce.
5. NASA/William Anders.
7. © 2012 Eames Office, LLC (www.eamesoffice.com).
12. NASA.
14. CDC/Dr. Frederick Murphy.
15. © (220546) CONACULTA.INAH.SINAFO.FN.MÉXICO.
16. © Contemporary Print from Original Negative by Ansel Adams, UCR/California Museum of Photography, Sweeney/Rubin Ansel Adams FIAT LUX Collection, University of California, Riverside.
17. © 2006 Zac Wolf.
18. Courtesy of George Eastman House, International Museum of Photography and Film.
19. Courtesy of Special Collections and Archives, University of California, Irvine Libraries. University Communications Photographs. AS-061.
20. Courtesy of Bain News Service, Print Collection.
21. Los Alamos National Laboratory.
22. Courtesy of Naval History and Heritage Command Photographic Department.
24. Photograph by Saša Šantić.
27. National Astronomy and Ionosphere Center.
28. © 2009 Jakub Halun.
29. Courtesy of General Research Division, The New York Public Library, Astor, Lenox, and Tilden Foundations.
30. © 2010 Ai Weiwei.
31. © 2007 Paul Synnott.
32. © 2011 Brett Elmer.
34. © 2010 Patrick Sanford.
36. © 2010 Nanex, LLC.
37. Courtesy The Bancroft Library, University of California, Berkeley.
38. © 2009 Matt Baer.
40. © 2011 Farm Sanctuary.
43. © Courtesy of AFMA.
45. © 2008 Tim Jones.

46. © 2008 CloneSafety.org.
47. NASA.
50. NASA.
52. Ohio Historical Society.
53. Courtesy of the U.S. Bureau of Reclamation.
57. © 2009 Simon Chetrit.
58. Courtesy of Alex Wellerstein.
62. © 2010 Roland Schmid.
63. National Archives of the UK, ref. HO382/3.
64. National Archives photo no. 210-G-E726.
65. © 2011 Essam Sharaf.
66. © Captain America and TM Marvel and Subs. Used with Permission.
68. Nat Farbman/Time & Life Pictures/Getty Images.
69. © Kirill Kay/National Geographic Stock.
70. Courtesy of the Yale Babylonian Collection .
74. From Maximum Power: The Ideas and Applications of H.T. Odum, edited by Charles A. Hall, © 1995 University Press of Colorado.
81. Timothy O'Sullivan/Time & Life Pictures/Getty Images.
82. Bundesarchiv, Bild 183-1987-0704-077/CC-BY-SA.
83. © 2008 Morningfrost.
84. © 2008 Lukas Kurtz. www.LuxTonnerre.com (Shootme@LuxTonnerre.com).
86. National Nuclear Security Administration.
87. Courtesy of the International Campaign for Justice in Bhopal.
88. © Nancy Hreha.
89. © 2012 F. Rudolf Turner, Ph.D.
90. © Fritz Bernhard. © 2007 Foundation of the Works of C.G. Jung, Zurich.
91. Courtesy of Al Jazeera English.
92. © Hans Hinz / ARTOTHEK.
93. Glacier National Park Archives.
94. U.S. Geological Survey.
96. National Oceanic and Atmospheric Administration/Department of Commerce.
98. National Oceanic and Atmospheric Administration/Department of Commerce.
99. © 2010 Noor Behram.

ADDITIONAL IMAGE CREDITS

Page vi: © 2012 EchoStar Corporation.
Page vii: © 2012 EchoStar Corporation.
Page xiv: © Trevor Paglen.
Page 5: © Trevor Paglen.
Page 11: © Hans Hinz / ARTOTHEK.
Page 17 (top): Courtesy of Pioneer Project, ARC, and NASA.
Page 17 (bottom): Courtesy of NASA.

Page 18: Courtesy of the U.S. Department of Energy.
Page 19 (top): Courtesy of the U.S. Department of Energy.
Page 19 (bottom): Courtesy of the U.S. Department of Energy.
Page 178: © Trevor Paglen.
Page 180: © Trevor Paglen.
Page 184: © Trevor Paglen.